Pgs. 1
1:
2
39
41
228-229

■ *FUNDAMENTALS OF DESIGNING FOR CORROSION CONTROL—* *A Corrosion Aid for the Designer*

R. JAMES LANDRUM

Published by

National Association of Corrosion Engineers
1440 South Creek Drive
Houston, Texas 77084

Library of Congress Catalog Card Number: 88-61244
ISBN 0-915567-34-2

Neither the National Association of Corrosion Engineers, its officers, directors, nor members thereof accept any responsibility for the use of the methods and materials discussed herein. No authorization is implied concerning the use of patented or copyrighted material. The information is advisory only and the use of the materials and methods is solely at the risk of the user.

Printed in the United States. All rights reserved. Reproduction of contents in whole or part or transfer into electronic or photographic storage without permission of copyright owner is expressly forbidden.

Copyright 1989
National Association of Corrosion Engineers

First printing
1989

Second printing
August, 1992

Third printing
September, 2000

Fourth printing
April, 2007

DEDICATION

To my wife Eileen
Without her love, loyalty, intelligence, and patience, I would never have accomplished much, including writing this book.

☐ ABOUT THE AUTHOR

R. J. Landrum is a materials engineer with over 38 years of experience in the field. He has worked with many design groups and industrial plants to improve the corrosion resistance of process equipment and structures in such diverse industries as chemical, paint, oil, agricultural chemicals, and atomic energy.

The author graduated from the University of Michigan (Ann Arbor, Michigan) with a Bachelor's of Science degree in Chemical Engineering. He was a member of Beta Theta Pi and Alpha Chi Sigma fraternities and was a three-year letterman on the wrestling team. Prior to his retirement, he was employed in the engineering department of E. I. du Pont de Nemours & Co., Inc., (Wilmington, Delaware) for 28 years, holding various supervisory and technical positions including Senior Consultant and Head of Applied Materials Engineering, where he managed a group of consultants.

Landrum has represented Du Pont in the American Welding Society (AWS) (Miami, Florida) for 25 years. During that time, he was chairman of several subcommittees for the Piping Committee, including "Practices and Procedures for Welding Austenitic Stainless Steel Piping and Tubing." His interest in welding is reflected in the chapters of this book dealing with the effect of welding on corrosion resistance (Chapters 3 and 4). In addition to membership in the AWS, Landrum has held memberships in the ASM International and the National Association of Corrosion Engineers (NACE).

Before joining Du Pont, the author was Chief Metallurgist at Una Welding, Inc. in Cleveland, Ohio; Chief Chemical Engineer for Walter Kidde and Co. in Belleville, New Jersey; and Materials Engineer at Kellex Corp. in Oak Ridge, Tennessee.

Landrum has lectured frequently in the United States and several times in Europe on the subject of designing for corrosion control. The author has written a number of technical articles dealing with corrosion. He has taught courses in metallurgy for the University of Delaware Extension, chaired a continuing education course in Modern Materials Engineering Technology at Du Pont (Wilmington, Delaware) for 9 years, and holds a patent on an apparatus for erosion-corrosion testing, which is now the basis for NACE Standard TM0270-70 ("Method of Conducting Controlled Velocity Laboratory Corrosion Tests"). He is a licensed professional engineer in Delaware and is an independent consultant in materials engineering.

■ CONTENTS

Preface
■ 1 — UNDERSTANDING CORROSION 1
   ☐ 1.1 Losses Resulting from Corrosion 1
      1.2 Designer's Role in Controlling Corrosion 2
      1.3 Equipment Service Life Factors 11
      1.4 Corrosion 13
      1.5 Frequency of Corrosion Failures 24

■ 2 — SELECTION OF MATERIALS OF CONSTRUCTION 27
   ☐ 2.1 Selection by Laboratory or Pilot Plant Testing 27
      2.2 Preparation and Assembly of Corrosion Test Specimens 27
      2.3 Corrosion Rate Determinations and Allowances................42
      2.4 Corrosion Rate Classification 46

■ 3 — DESIGN SOLUTIONS TO CORROSION PROBLEMS
         BASED ON TYPE OF CORROSION 49
   ☐ 3.1 Crevice Corrosion 49
      3.2 Galvanic Corrosion 67
      3.3 End Grain Attack 81
      3.4 Erosion-Corrosion 86
      3.5 Fretting Corrosion 101
      3.6 Pitting 103
      3.7 Microbiologically Influenced Corrosion............................107
      3.8 Stress Corrosion Cracking 115
      3.9 Hydrogen Induced Cracking, Hydrogen Embrittlement,
           and Liquid Metal Cracking 137
      3.10 Corrosion Fatigue..140
      3.11 Intergranular Corrosion 144
      3.12 Dealloying 155
      3.13 Hydrogen Grooving 158

- **4 — DESIGN SOLUTIONS TO CORROSION PROBLEMS BASED ON FABRICATION TECHNIQUES AND ENVIRONMENTAL FACTORS** 167
  - 4.1 Welding 167
    - 4.2 Heat Treatment 187
    - 4.3 Temperature 193
    - 4.4 Free Drainage 197
    - 4.5 Environmental Corrosion 202
    - 4.6 Fresh Water 205
    - 4.7 Automotive Corrosion 211

- **5 — SPECIFICATIONS AND GUIDES** 217
  - 5.1 Specification Writing for Maximum Corrosion Resistance 217
    - 5.2 Stainless Steels 230
    - 5.3 Nickel and Nickel Alloys 239
    - 5.4 Aluminum and Aluminum Alloys 246
    - 5.5 Carbon and Low Alloy Steels 248
    - 5.6 Copper and Copper Alloys 254
    - 5.7 Titanium and Titanium Alloys 257
    - 5.8 Unified Numbering System (UNS) 260
    - 5.9 Overall Materials Guide (OMG) 261

- **6 — CONTROL TECHNIQUES** 267
  - 6.1 Quality Control 267
    - 6.2 Corrosion Monitoring 286
    - 6.3 Cathodic Protection 293
    - 6.4 Anodic Protection 303
    - 6.5 Inhibitors 310
    - 6.6 Protective Coatings 314

References 333
Appendix 337
Index 339
Cross Index 353

☐ PREFACE

Corrosion problems can be instigated during the design stage of a project because of the designer's lack of corrosion awareness. He may not realize that the corrosion resistance of the equipment he is designing may be reduced by an unwitting decision or action on his part.

If potential corrosion problems are not resolved during the design phase, they will inevitably arise later in the forms of corroded equipment and unexpected shutdowns. In such cases, high costs for repair or replacement of equipment as well as the cost incurred by the suspension of production during shutdowns will result. These problems can be averted if the designer weighs *corrosion resistance* as heavily as, for example, strength of materials.

As Mars Fontana stated in the Campbell Memorial Lecture (at the NACE 1970 Annual Conference), "virtually all corrosion failures result from carelessness on the part of the user or *poor choice of material or configuration by the designer.*"

The purpose of this book is to enhance the designer's knowledge of what he can do to reduce the very high cost of failures caused by corrosion. This book is not just limited to mechanical design of equipment, but it also encompasses corrosion testing and selection of materials of construction; specification writing; quality control and required testing determination; corrosion monitoring; and corrosion control techniques.

To present the wide range of corrosion control options available to the designer within the space limitations of this book, the information has been condensed. The fundamentals of each option are briefly discussed, and references are noted for more detailed treatises on most of the subjects covered.

This book is primarily concerned with metals and alloys; whereas, nonmetals, except for plastic pipe, are not covered.

The many cases of equipment failures and examples of good corrosion design practice found in this book (those which are not referenced) are drawn from my own experience in materials engineering over the years.

R. James Landrum
Wilmington, Delaware
January 2, 1989

☐ ACKNOWLEDGMENTS

I wish to acknowledge the assistance I have received in writing this book. R. E. Tatnall and G. Kobrin furnished invaluable information and pictures on microbiologically influenced corrosion; J. C. Bovankovich was very helpful in his field of corrosion monitoring; N. E. Hamner aided me with anodic protection; and S. K. Brubaker was helpful regarding sulfuric acid technology. My thanks also to T. F. Degnan for some of his case histories and to W. I. Pollock for initiating the writing of this book.

I am indebted to the Savannah River Plant for permission to use their pictures, to the American Welding Society for permission to use their drawings on welding symbols, and to the Society of Automotive Engineers (Warrendale, Pennsylvania) for permission to use their drawings on automobile corrosion principles.

R. James Landrum
Wilmington, Delaware

"TOM DEGNAN, THE CORROSION ENGINEER, WANTS TO SEE YOU ON A LITTLE MATTER!"

■ 1 — UNDERSTANDING CORROSION

☐ 1.1 Losses Resulting from Corrosion

The public worldwide is significantly affected by corrosion. Whether the individuals involved are engineers, businessmen, scientists, laboratory technicians, or average citizens, everyone pays the price in some manner for the ravages of corrosion. Annual losses resulting from corrosion in the United States, for example, are estimated to be many billions of dollars. The specific cost is debatable, since there is little consistency with regard to what is included in the calculation of the loss. Some losses resulting from corrosion, however, have been documented. For example:

1. $300 million are expended every year simply to replace 2,400,000 corroded hot water heaters.
2. It costs almost $40 million annually to maintain waterworks pump capacities just to compensate for the rust that is clogging up piping systems.
3. Experts from the Environmental Protection Agency (EPA) (Washington, DC) believe that leaking gasoline tanks present a widespread threat to the water supplies in the United States.[1] 2.3 million buried steel tanks are being attacked by corrosion. It is estimated that 25% of these tanks, or more than half a million, are leaking badly now as a result of corrosion and thus require replacement.
4. The chemical industry is particularly susceptible to losses resulting from corrosion because of the aggressive materials that are commonly handled in that industry. A survey made by one large chemical company in 1970 showed that its total estimated loss from corrosion for a 12-month period was $13.7 million. The annual loss caused by corroded heat exchangers alone in that company was $3 million in replacement and repair costs.
5. In the United States, 40% of the country's steel production is used to replace corroded parts and products.[2]

Unfortunately, the equipment replacement cost is not the only cost resulting from corrosion. These additional costs account for the difficulty in arriving at an accurate overall corrosion loss figure. Additional costs include the costs of the return of poor grade products from corrosion contamination, unscheduled plant shutdowns, loss of efficiency of heat transfer surfaces, and accidents resulting from corrosion.

One of these additional costs, unscheduled plant shutdowns, can affect not only the plant where the corrosion failure occurred, but may also shut down other plants dependent on the first plant for their materials. In anticipation of unscheduled shutdowns, plants often invest in either duplicate facilities or additional storage capacity. These costs are seldom credited to corrosion.

Throughout history, corrosion has been costly. In the era of Benjamin Franklin, for example, one manifestation was the "dry bellyache" with accompanying paralysis, which was mentioned by Franklin in a letter to a friend. This malady was actually caused by the ingestion of lead from corroded lead coil condensers used in making brandy. The problem became so widespread that the Massachusetts Legislature passed a law in the late 1700s that outlawed the use of lead in producing alcoholic beverages.

The designer at the drawing board, the process engineer, the chemical engineer, the process manager, the project manager, the vice president in charge of engineering, the company owner, etc., all have a vested interest in reducing corrosion costs.

In this book, the designer will be referenced frequently. Here, the term *designer* means anyone who has any responsibility for mitigating corrosion.

☐ 1.2 Designer's Role in Controlling Corrosion

The designer is in an effective position to lower the tremendous cost of corrosion if he recognizes that there could be a problem and he has the knowledge to act upon it. For example, he should:

1. Be aware of technical assistance that is available to him.
2. Have a well-defined course of action to follow to determine optimum materials.
3. Have the ability to calculate the most economical selection from a number of corrosion control methods that have been determined to be able to perform well for a given application.

Using Professional Consultants

Many large companies, particularly in the chemical industry, have materials engineering groups comprising trained engineers who work directly with designers and company plants to help reduce corrosion. Figure 1.1 displays an organizational structure used for one such group.

Materials engineers have intimate knowledge of the processes and corrosion problems in their assigned plants. They are then available to the designers for consultation on any design problem. The designer, in turn, must have a basic knowledge of corrosion to recognize when a problem may exist and when to consult a materials engineering expert. The fundamental obstacle is getting corrosion problems confronted while the process is still at the "drawing board" stage, rather than after the plant has been built.

Many smaller companies do not have the luxury of in-house materials engineering groups. Therefore, designers, in many instances, may have to rely on vendors of engineering materials for advice in the selection of materials of construction, design details, specifications, etc. There are many materials suppliers who provide beneficial information on the materials they manufacture, having conducted many corrosion and mechanical tests on specific products. They also have knowledge about how well their product has performed in the field. However, caution should be exercised by the designer in acting solely upon the vendor's recommendations. To be good salesmen, vendors have to be "sold" on their own products.

For example, the paint salesman may want to paint over everything, while the stainless steel vendor may want to make everything out of stainless steel. As a solution to this problem and to avoid discouraging

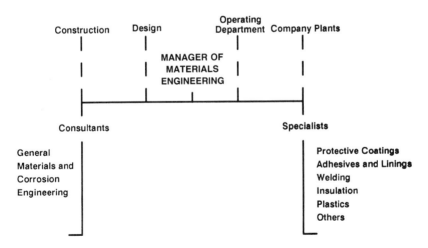

FIGURE 1.1 — *Organizational structure chart for a materials engineering group.*

designers from using vendors (as vendors can render useful services, it is wise to have the designer follow an established procedure so that he will have no doubt about which is the better alternative, in this case, paint or stainless steel.

Professional corrosion engineering assistance is available from a variety of sources. The designer can inquire through his own technical society or peruse advertisements in technical journals dealing with metals and corrosion for assistance. For Instance, in *Materials Performance*, a monthly journal published by the National Association of Corrosion Engineers (NACE),[1] there is a section that lists available corrosion engineering assistance.

Optimum Materials Determination

The designer should study a variety of materials through pertinent tests and review of past experience, and then he should make a selection based on the material's ability to do the job, safety, and economics. Candidate materials recommended by vendors can be included in the various materials to be tested. The designer is responsible for matching his unique materials problems with the best materials and design, which requires evaluating pertinent materials on the market to attain the "best fit." The best fit is not necessarily the most or the least expensive material. For example, perhaps all corrosion problems for a specific application would disappear if all equipment were made of platinum. Of course, a plant made of platinum would most likely be too expensive to be justified. The designer should use the optimum material for a specific application. *Optimum* simply means the least expensive material that will do an adequate and safe job.

Regardless of what was stated above, platinum is used sometimes as a construction material. In one plant, not only was platinum clad columbium the most inexpensive material for a critical electrolytic dissolver part, but it was the only material that could do the job.

Selecting the Most Economical Corrosion Control Method

There are many alternatives in solving corrosion problems, as this book will later discuss. Whether the choice is corrosion-resistant materials or less expensive materials using electrochemical techniques, coatings, inhibition, etc., for protection, the basis for the selection requires the recognition and appraisal of economic factors, as well as an

---

[1]NACE, Houston, Texas. (All association names and addresses are given in the Appendix.)

understanding of corrosion technology. Dillon emphasized that corrosion is basically an economic problem.[3] For these reasons, the designer should know how to calculate annual costs of various alternatives for corrosion control.

A cost appraisal method should be used that will determine the one corrosion control method that offers the greatest economic advantage. The appraisal should consider all economic aspects such as cost, interest rate, life, and depreciation schedule.[3] The cost of an anti-corrosion alternative is not only the original installed cost, but it is all costs, including operating, maintenance, overhead, and various interim costs.

Interest is simply the value of money, whether the money is borrowed or company funds that otherwise would draw interest are being used.

The *life of the equipment* is an estimate of the anticipated life of an alternative. This may be determined through past experience, laboratory tests, or pilot plant tests.

The *depreciation schedule* of an alternative is the depreciation write-off used in calculating taxes. The following methods are used:

— The *straight line* method is one in which the rate of depreciation is constant for each year of the equipment's life.
— The *sum-of-the-digits* method incorporates faster the early year's write-offs, as do the *various declining balance* methods.

Recognizing the need for an overall approach to aid in selection between anti-corrosion alternates, NACE has published Standard Recommended Practice RP0272-72 ("Direct Calculation of Corrosion Cost Control Methods").[4] This document should be valuable to the designer since it considers all of the factors discussed above, including equipment installation costs, maintenance expenses, depreciation costs, cost of money, rate of return after taxes, and tax rates. Formulas for determining overall annual costs of various alternates have been developed that reflect these factors. The symbols used are:

A = Annual overall costs
d = Tax and depreciation factor
C = Capital outlay
K = Annual expense
i = Interest rate as a decimal
r = Rate of return after taxes as a decimal

F = Variant form of capital recovery factor in standard financial tables
t = Tax rate as a decimal

There are two calculation methods advocated in this NACE Standard RP0272-72. They are termed *annual costs* (A) and *present worth after taxes* (PWAT). The annual costs method appears to be the easier method to understand and use and therefore is used here in the annual cost calculations of the examples. (PWAT can be calculated from A or *vice versa*.) The basic formula for the annual calculation is:

$$A = -C^{(2)} \times d_n \times r \times F_n \qquad (1.1)$$

The derivation of this formula may be found in NACE Standard RP0272-72. Table 1.1 shows values for d and F based on:[3]

— Sum-of-digit depreciation,
— Tax rate of 48%, and
— Money worth 10% after taxes.

As an example of how economical appraisals can be made, suppose the designer is considering two pipeline designs:

1. A carbon steel pipeline can be installed for a cost of $30,000. The yearly maintenance expenses will be $4200. The expected life of the installation based on past experience is 5 years.
2. A stainless steel pipeline can be installed for $82,500. There will be no yearly maintenance expense. The estimated life of the installation is 10 years.

Calculations for a Steel Pipeline

$$A_{steel} = -C \times d_5 \times r \times F_5 \qquad (1.2)$$

Where: $-C = -\$30{,}000$
$d_5 = 0.617^{(4)}$
$r = 0.1$
$F = 2.637^{(4)}$

---

[2] The capital outlay has a negative sign because it represents a negative cash flow.
[3] For cases where variables other than those shown above are used, an accountant should amend the figures in Table 1.1.
[4] From Table 1.1.

TABLE 1.1
Tax and Depreciation Factors Used in Calculating Economic Appraisals of Corrosion Control Measures[1]

Years (n)	$(1+r)^n$	d_n	F_n
1	1.100	0.520	11.000
2	1.210	0.556	5.762
3	1.331	0.583	4.021
4	1.464	0.603	3.155
5	1.611	0.617	2.637
6	1.771	0.627	2.297
7	1.948	0.633	2.055
8	2.145	0.636	1.873
9	2.353	0.639	1.736
10	2.595	0.639	1.627
11	2.855	0.640	1.539

[1]Abbreviated from NACE Standard RP02-72 ("Direct Calculation of Corrosion Cost Control Methods"). (n) is the life of the project in years; (r) is the rate of return (in dollars) after taxes as a decimal; (d) is the tax depreciation factor; and (F) is the variant form of capital recovery factor in standard financial tables.

Then: A_{steel} = −$30,000 x 0.617 x 0.1 x 2.637
= −$4881

Since maintenance is an *annual expense*, both Factors d and F are effective for *one* year. Therefore,

$$A_{\text{maintenance expense}} = -K \times d_1 \times r \times \frac{F_1}{(1+r)^1} \quad (1.3)$$

Where: −K = −$4200
d_1 = 0.520[5]
r = 0.1
F_1 = 11[5]
$(1 + r)^1$ = 1.1[5]

Then:

$$A_{\text{maintenance expense}} = -\$4200 \times 0.520 \times 0.1 \times \frac{11}{1.1}$$

[5]From Table 1.1.

$$= -\$4200 \times 0.520 \times 0.1 \times 10$$
$$= -\$2184$$

Therefore, the *Total* Annual Cost for steel pipeline =

$$= A_{steel} + A_{maintenance\ expense}$$
$$= -\$4881 - \$2184$$
$$= -\$7065$$

Calculations for Stainless Steel Pipeline

$$A_{stainless} = -C \times d_{10} \times r \times F_{10} \quad (1.4)$$

Where: $-C = \$82,500$
$d_{10} = 0.639^{(6)}$
$r = 0.1$
$F_{10} = 1.627^{(6)}$

Then: $A_{stainless} = -\$82,500 \times 0.639 \times 0.1 \times 1.627$
$= -\$8577$ (since there is no maintenance cost, this is the *Total* Annual Cost)

Conclusion: $A_{stainless} - A_{steel} = \$8577 - \$7065$
$= \$1,512$

The best selection would be the carbon steel pipeline, because the annual cost is less, resulting in an annual cost saving of $1512. Even though the stainless steel pipeline would last twice as long, its higher original cost must be written off in ten years, while the maintenance cost of the carbon steel pipeline, being an expense, can be written off each year.

Mistakes have been made while quickly estimating costs of alternate materials selections because important factors such as taxes and cost of money were not considered or thought to be significant. For instance, suppose the cost of the carbon steel pipeline was calculated for ten years, as follows:

Installed cost plus one replacement
($30,000 \times 2$) = $60,000
Cost of Maintenance 4200×10 = $42,000
+ _____

Total Cost = $102,000

---

[6] From Table 1.1

Since the stainless steel costs $82,500 to install, has no maintenance costs and lasts 10 years, $82,500 is obviously the *total* cost for 10 years.

Conclusion: $A_{steel} - A_{stainless} = \$102,000 - \$82,500$
$= \$19,500$

An erroneous conclusion would indicate that the *stainless steel* was the least expensive and would save $19,500 in ten years or $1950 per year.

Another question that could be answered by this economic appraisal method is:

> How much of a reduction in the installation cost of the stainless steel pipeline would have to be made before the annual cost of the stainless steel pipeline would be as low as that of the carbon steel pipeline?

The annual cost of the stainless steel pipeline is set to be the same as the carbon steel pipeline, as follows:

$A_{stainless} = A_{steel} = \$7065$

$$A_{stainless} = -C \times d_{10} \times r \times F_{10} \tag{1.5}$$

Where:
  $-C$ is to be determined
  $d_{10} = 0.639$[7]
  $r = 0.1$
  $F_{10} = 1.627$[7]

Divide both sides of the equation by $d_{10} \times r \times F_{10}$

Then: $\quad -C = \dfrac{A_{steel}/A_{stainless}}{d_{10} \times r \times F_{10}}$

Therefore: $\quad -C = \dfrac{7065}{0.639 \times 0.1 \times 1.627}$

---

[7]From Table 1.1.

$$= \frac{7065}{0.1040}$$

$$-C = \$67,933$$

Conclusion: The installed cost of the stainless steel pipeline would have to be reduced by $14,567 ($82,500 − $67,933) before its annual cost would be as low as that of the carbon steel pipeline. (See Table 1.2 for a listing of these three examples.)

TABLE 1.2
Cost Considerations in Selecting a Pipeline

	Carbon Steel	Stainless Steel	
Installed Cost	$30,000	$82,500	$67,933[1]
Life (in years)	5	10	10
Yearly Maintenance	$4200	0	0
Annual Cost	$4881	$8577	$7065
Annual Maintenance Cost	$2184	0	0
Total Annual Cost	$7065	$8577	$7065

[1] The installed cost of the stainless steel pipeline to show an annual cost the same as that for a carbon steel pipeline.

Nothing is more important to the designer than to have accurate corrosion cost information. The following two examples may help emphasize this:

Example 1 — A materials engineer visited an acid manufacturing plant and asked the plant personnel if there were any corrosion problems. He was assured that the plant had none. Later, the materials engineer saw a stack of pipe all cut to to the same length in the center of the plant and inquired about them. Again, he was assured that this corrosion problem had been solved. They said that, periodically, when an installed length of pipe corroded through, it was quickly replaced with a new piece already cut to length; the time required for the replacement only required about 1 hour of work. As a result of the engineer's visit, a very expensive piece of corrosion-resistant pipe replaced the steel pipe and has lasted for many years with no replacement needed.

Example 2 — The upper management of an oil company decided that a better material of construction was desired for five heat exchangers in an atmospheric topping unit that was being built.[5] Stainless steel was thought to be much more effective than the normally used, long-lasting carbon steel based on the belief that the more expensive material would be the more corrosion resistant. The stainless steel heat exchanger tubes failed by chloride stress corrosion cracking during the first week of service.

☐ 1.3 Equipment Service Life Factors

The designer is confronted with three primary concerns regarding materials of construction as he begins the design:

1. Resistance to stress;
2. Resistance to wear; and
3. Resistance to corrosion from process and atmospheric conditions.

The first two concerns, involving mechanical properties of materials, such as tensile strength, yield strength, ductility, fatique strength, wear resistance, etc., are, of course, very important; designers are usually well aware of them. However, designers may not be aware of possible corrosion problems.

Corrosion resistance should be as important during the design stage as other operational conditions, such as velocity, temperature, pressure, etc. Mechanical stress in structures can usually be rather precisely predicted; however, the prediction of the destructive effect of corrosion cannot be predicted as accurately, particularly in new processes.

The selection of the proper material of construction is an important part of the designer's job and is the one factor that is generally emphasized. However, consider all of the following factors that influence the equipment's service life:[6]

—Selection of materials of construction
—Design details
—Specification of materials
—Fabrication and inspection
—Process operation
—Maintenance

In other words, for the best equipment or structural design, the **materials of construction** must be carefully selected from a corrosion-resistance standpoint. The **design details** should preserve the corrosion resistance of the materials. Concise and clearly written specifications should be provided to the supplier to ensure that the material needed is accurately ordered. The equipment should be **fabricated** properly and adequately **inspected** to prove compliance with the specifications. The equipment must be **operated properly**. (This last item is sometimes overlooked; plants may change a process without sufficient regard to the effect of the process change on the construction materials.) Lastly, the equipment must be **maintained** properly. All of these factors must be considered by the designer to ensure the long life of the equipment he designs. When corrosion failures occur, the selection of the materials of construction involved is usually faulted. However, in a large number of cases, failure actually occurred because of other factors. The following examples may clarify this point:

Example 1 — A materials engineer was called upon by a plant to determine better materials of construction for a critical stainless steel heat exchanger that corroded so severely that according to the plant, it had had to be replaced four times. The rejected heat exchangers had been disposed of in the scrap yard. Upon examination, the engineer found that the tubes of the rejected heat exchangers were severely clogged with corrosion products. After scraping these away, he found a bright stainless steel tube wall underneath, indicating that no corrosion had actually occurred there. Further investigation disclosed that equipment made of carbon steel upstream from the heat exchanger was unexpectedly disintegrating by corrosion. These corrosion products were discharged into the process stream. The heat exchanger afforded the most restricted area in the system, and the corrosion products collected there and clogged up the tubes. The corroding carbon steel equipment was subsequently replaced with a more resistant material of construction, and the plant operated on cleaned exchangers from the scrap yard until the plant shut down years later.

In this case, not only was the wrong material faulted for the failure, but the corrosion problem had not been properly defined. It has been said that the definition of a problem is often the solution to the problem; that axiom certainly proved true here.

Example 2 — A materials engineer was called to a plant because a new

AA[8] 3003 aluminum alloy tank, which stored 95% nitric acid (used for a nitriding process step), was leaking at all of the welds, while the rest of the tank appeared unaffected. The investigation disclosed that the welds were made using an AA 4043 aluminum filler wire, which has a high silicon content (5%) and is very susceptible to attack by 95% nitric acid. The original order had failed to specify the type of filler metal; the fabricator had thus used the filler metal that was easiest for his welders to use. This corrosion failure, resulting simply from a poorly written specification, caused an unscheduled plant shutdown. As a consequence, all of the welds had to be ground out of the tank and rewelded with the proper filler wire alloy, AA 1100 aluminum. This vessel has now been in service for many years with no evidence of corrosion.

Example 3 — Shells made of 18% nickel maraging steel were ordered for special paper manufacturing rolls to replace AISI[9] H-11 tool steel shells that had given good service. These shells were heated by internal electrical resistors. The heat transfer medium inside the roll was a 50:50 lead to bismuth alloy. The maraging steel was selected because of its lack of distortion during heat treatment. This lack of distortion simplified fabrication, but a poor assumption was made that because the transfer media would not affect carbon steel or austenitic stainless steel, the medium would not affect the new alloy either. Fortunately, a test on stressed specimens was conducted before the shells were placed in service. In this test, the maraging steel failed by catastrophic cracking in less than 30 hours. The shells were consequently scrapped and new shells were made of the old material.

The six factors influencing the equipment's service life should be kept in mind by the designer. Various aspects of these factors will be dealt with in greater detail in subsequent chapters.

☐ 1.4 Corrosion

Corrosion can be broadly defined as material deterioration by chemical or electrochemical attack. Obviously, an understanding of the basic corrosion mechanisms is essential in designing equipment for corrosive service. The corrosion process is complex, and for that reason, only the

---

[8] Aluminum Association (AA), Washington, DC; see Appendix for complete address.
[9] American Iron and Steel Institute (AISI), Washington, DC; see Appendix for complete address.

highlights will be discussed here. For those designers who wish to investigate this subject in more detail, References 7 and 8 are recommended.

The Corrosion Cell

The manner in which corrosion progresses can be learned by studying the basic corrosion cell. This cell comprises four major components:

- Electrolyte
- Anode
- Cathode
- Electronic circuit

If two different metals, copper and zinc for example, are connected electrically and immersed in an electrolyte, an electrical current will be generated. This will result in corrosion of the zinc, while the copper remains unaffected. (See Figure 1.2.) The zinc acts as the anode, and the copper acts as the cathode.

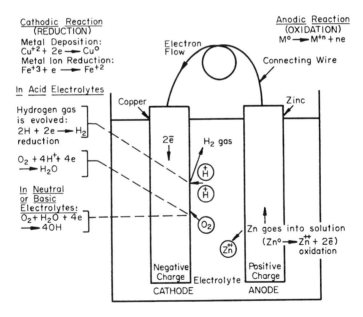

FIGURE 1.2 — *The corrosion mechanism. Note: to avoid confusion, the electrical current flows in the opposite direction of the electron flow.*

This sequence is an electrochemical reaction, which is defined as a chemical reaction involving the transfer of electrons, which, in turn, involves oxidation and reduction.

The electrolyte. A solution that is able to conduct electricity is called an *electrolyte*. The ability to conduct electricity is a result of the presence of ions in the solution. Ions are positively or negatively charged atoms in solution. For example, water dissociates to a small extent, producing equal quantities of hydrogen and hydroxyl ions:

$$HOH \rightarrow H + OH \quad (1.6)$$

Oxygen is often dissolved in electrolytes and has an aggressive influence on the corrosion of metals. The electrolyte may be any solution, such as seawater, rainwater, acids, bases, etc. In Figure 1.2, two cases are reviewed where the electrolyte is (1) acidic and (2) neutral or basic.

The anode. Zinc, which has a valence of 0, goes into solution in the electrolyte as a positively charged zinc ion. The zinc now has a valence of +2; thus, the zinc has been oxidized. As it goes into solution, the zinc leaves behind two negative charges in the form of electrons. These electrons then flow through the metal and through the connecting wire. The following equation expresses this anodic reaction:

$$Zn \rightarrow Zn^{+2} + 2\,e \quad (1.7)$$

Where e = electrons.

The anodic reactions occurring during corrosion of all metals can be expressed as:

$$M \rightarrow M^{+n} \times ne \quad (1.8)$$

where M is the corroding metal that oxidizes to an ion with a valence charge of +n and a release of n electrons. This equation applies to all corrosion reactions, regardless of the nature of the electrolyte or the corrodent.

The cathode. At the cathode, a reaction occurs that consumes the electrons flowing from the anode and consequently restores the electrical balance to the system. In the example in Figure 1.2, consider an acid electrolyte. Since the acid electrolyte, by definition, has many hydrogen ions (the lower the pH, the greater the quantities of hydrogen ions are), these hydrogen ions are available and are reduced at the cathode and hydrogen gas will form (valence change from +1 per hydrogen ion to 0 for hydrogen gas):

$$2H^{+1} \times 2e \rightarrow H_2^0 \tag{1.9}$$

Also in acid solutions, oxygen can be reduced to water:

$$O_2 + 4H^+ + 4e \rightarrow 2H_2O \tag{1.10}$$

If the electrolyte is neutral or basic, the cathodic reaction will be different. There must be dissolved oxygen in the electrolyte to allow the reaction to continue. At the anode, the reaction will be the same as it was for an acid solution; however, at the cathode, no hydrogen gas will be formed, but the dissolved oxygen will be reduced as it reacts as follows:

$$O_2 + 2H_2O + 4e \rightarrow 4OH \tag{1.11}$$

Oxygen is the key to this reaction. If oxygen is removed from neutral or alkaline water (such as boiler water), the corrosive action will be halted.

For this reason, sodium sulfite or hydrazine is added to boiler water to remove the dissolved oxygen. Deaeration of the incoming boiler water achieves the same result.

Two other cathodic reactions are associated with the corrosion of metals:

$$Fe^{+3} + e \rightarrow Fe^{+2} \tag{1.12}$$

Or metal deposition can occur at the cathode:

$$1Cu^{2+} + 2e \rightarrow Cu^0 \tag{1.13}$$

The electronic circuit. All of the reactions shown in Equations (1.7) through (1.13) depict the consumption of electrons at the cathode; as long as this action continues, electric current will flow and corrosion will continue. Almost all metallic corrosion can be demonstrated by one or several of these reactions individually or combined. In every case, an oxidation occurs at the anode and a reduction occurs at the cathode. For corrosion to occur in metals, there must be a release of electrons at the anodic surface where corrosion is found, and the electrons must be consumed at the cathode. These reactions occur simultaneously and at equivalent rates.

What a Designer Can Do to Reduce or Stop Corrosion

A practical aspect in understanding the corrosion cell is the conception of how corrosion can be reduced or stopped completely. Referring again to Figure 1.2:

1. If the wire between the anode and the cathode is cut, the corrosion current (caused by galvanic acceleration) stops. This is the basis for insulating one member of a galvanic couple from another. In effect, the electrical connection between the two is severed.
2. If either the copper or the zinc electrode is covered with an impervious coating, the electrical path is disrupted and corrosion ceases. This, of course, is one reason for applying protective coatings.
3. If an outside current (DC) is impressed on the cell in the opposite direction that is equal or greater than the corrosion current, corrosion is reduced or stopped. This is the basis for cathodic protection.
4. If the electrolyte were eliminated, there would be no way of transporting ions; thus, corrosion ceases. This is the primary reason for keeping metals dry.

As indicated above, each of these methods for reducing or halting corrosion is commonly used in industry. These control methods will be discussed in greater detail in later chapters.

Corrosion Cell Creation

As shown in Figure 1.2, a corrosive cell is formed because two dissimilar metals, e.g., zinc and copper, in an electrolyte form an anode and a cathode. Many other conditions can form corrosion cells as well. Some of these conditions are:

Difference in structure. Corrosion is a possibility when metals exhibit areas where there are: (1) differences in structure, such as grain boundaries versus metal matrix, (2) differences in grain orientation, and (3) second-phase constituents versus adjacent solid solutions.

Difference in composition caused by impurities and defects. As an example, sulfide inclusions in steel can set up anodic areas that can establish corrosion cells with surrounding cathodic areas. It has been demonstrated that as the purity of a metal is increased, the tendency for that metal to corrode is reduced.

Difference in aeration. If one area of a solution is aerated more than another area that has a lesser oxygen content, the more aerated area will become the cathode. The area with the lower oxygen content will become the anode. Such a cell is called an *oxygen concentration cell* and will be discussed in more detail in Chapter 3.1.

Differences in metal ion concentration. Where the concentration of a metal ion in a solution is different in one area compared to that of an adjacent area, a corrosion cell can be established. The area of the lower metal ion concentration becomes the anode, and the area of the higher ion concentration becomes the cathode. Such a cell is termed a *metal ion concentration cell* and will also be discussed in a Chapter 3.1.

Differences caused by thermal treatment. Any variation in the homogeneity of metals caused by heat treatment effects can cause cells to form. For instance, quenching effects that distinguish one area from a neighboring area can produce anodic and cathodic areas.

Difference in strain. Where there are adjacent areas of differing strain, corrosion cells can form. For instance, consider a piece of metal that is bent and placed in an electrolyte. The strained area at the bend can form an anode and be corroded, and the unstrained part can become the cathode and will not corrode.

Corrosion of Steel in Water and Acid

Water. Because of the inhomogeneity of many metals' structure and/or composition, large numbers of anodic and cathodic areas are formed that can cause corrosion. For instance, when steel is immersed in water containing dissolved oxygen and consequently rusting occurs, the following events take place.

At the anode, the iron is oxidized to ferric iron ions, Fe^{+++}. At the time the ferric ions go into solution, three electrons per ion remain in the metal and flow to the cathode. At the cathode, the electrons then reduce the dissolved oxygen. The iron ions combine with water and oxygen to form ferric hydroxide, which is insoluble and precipitates out of solution. This reaction is shown in the following equation:

$$4\ Fe\ +\ 6H_2O\ +\ 3O_2 \rightarrow 4\ Fe\ (OH)_3 \downarrow \qquad (1.14)$$

During drying periods in the atmosphere, the ferric hydroxide is dehydrated, leaving the familiar reddish-brown rust behind. This reaction is shown in the equation below:

$$2\ Fe(OH)_3 \rightarrow Fe_2O_3 + 3H_2O \uparrow \tag{1.15}$$

(Because of other elements being present, rust is actually a combination of mixed iron oxides, hydroxides, and other compounds, such as carbonates.)

Acid. When steel is immersed in sulfuric acid (below about 78%), corrosion is rapid and the steel readily dissolves. Actually, the corrosion mechanism is much the same as that for rusting, except that ferrous sulfate is formed instead of ferric hydroxide. Figure 1.3 shows that the iron is oxidized at the anode, as was the case during rusting. However, at the cathode, the hydrogen ion is reduced to hydrogen atoms, which,

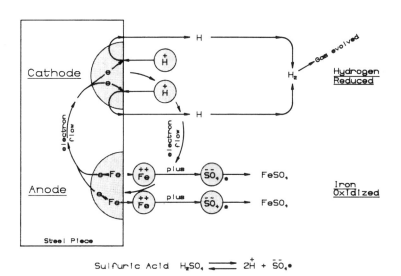

FIGURE 1.3 —*Steel dissolved in sulfuric acid.*

in turn, combine to form hydrogen gas.

Polarization

Process description. Corrosion action tends to slow down because of materials formed during the anodic and cathodic reactions. This is called *polarization*. It is important for the designer to be aware of this phenomena, since polarization determines, to a great extent, the corrosion rates of many metals and alloys. The conditions fostering polarization include formation of corrosion products and films and the concentration of unwanted species and lack of wanted species at the anodic and cathodic surfaces. These conditions *reduce* the severity of the corrosion action. Polarization can have a very great or small effect, depending upon the materials involved, the ratio of areas of the anode and cathode, and the specific environment involved.

There are several ways in which polarization can occur, which include:

1. When hydrogen ions are reduced at the cathode to hydrogen atoms (as previously described) and are not dispersed and stay in the cathode area, they prevent other hydrogen ions from reaching the cathode surface and reacting, thus retarding corrosion.

    (Conversely, *depolarizers* speed up corrosion. For instance, if air is present in a solution, the oxygen will combine with the hydrogen at the metal surfaces and thus remove the hydrogen from the area.)

2. If the two hydrogen atoms are slow in combining to produce one molecule of hydrogen gas or have not dissipated quickly from the cathode area, the corrosion action will be reduced.

3. The speed of electron transfer to the hydrogen ions at the metal surface, if slow, can also be a retardation factor. This is called *activation polarization*. The inherent rate depends on the metal or alloy involved, the temperature, and the hydrogen ion concentration of the corrodent.

Polarization curves for determining corrosion rates. As discussed previously, when corrosion occurs, an electrical current flows from the anode to the cathode. The greater the current, the greater the corrosion rate. If we could easily determine the magnitude of this current, we could calculate the corrosion rate using Faraday's laws. However, this is not very feasible because the corrosion current flows from a great number of small anodes on the surface of a metal to an equally great number of small cathodes. The corrosion rate, however, can be measured *indirectly* by determining anodic and/or cathodic polarization curves in a laboratory that has special equipment.

In general, this is done by exposing a metal specimen (whose corrosion rate in a specific environment is desired) to an external electrical circuit. The external circuit applies a potential (voltage) to the specimen, which changes the potential of the specimen to make it perform like an anode, with an inert electrode in the solution acting as the cathode. By applying many different potentials to the specimen and recording the resultant current, an *anodic polarization curve* can be constructed. By controlling the electrical potential, the specimen can also be made to act like a cathode and the inert electrode can be forced to perform as an anode. Recording the potentials and the resultant currents can produce a *cathodic polarization curve*. (Usually, only one curve is required for corrosion rate determination.)

After finding the potential of the freely corroding specimen (this determination is easy to make), its accompanying *current* may be determined from a *linear* polarization curve. Thus, by knowing the electrical current of the freely corroding specimen, the specimen's corrosion rate can be easily calculated.

Polarization and the Designer. The designer should have some knowledge of polarization because many corrosion rate determinations are being made using linear polarization curves. Alloy development is sometimes based on such curves. As will be discussed in Chapter 2.3 and 2.4, a corrosion test is no better than the accuracy with which the test duplicates actual process conditions. This should be remembered when a designer contemplates using linear polarization generated data for materials selection in a process that he is designing.

Corrosion monitoring equipment based on linear polarization is being used universally for remote corrosion testing of process streams. Chapter 6.2 will address this subject in more detail.

It is recommended that the designer consult Reference 9 for detailed information on corrosion rate determination from polarization data.

Passivity

Passivity is defined as the ability of certain metals and alloys to show low corrosion rates when they would be expected to corrode rapidly based on thermodynamics.

Role of oxygen. The tendency for a metal to become passivated is the property of the *anode* and will show up in the anodic polarization curves; however, whether or not a passive state is attained depends on the cathode characteristics. The nature of passivity is not well understood, but it is believed that *oxygen* (or an oxidizing agent) is responsible because it causes a very thin oxide film to be formed on the surface of some metals and alloys, which acts as a corrosion shield. The designer should not consider this film as similar to a coat of paint because when the metal is withdrawn from the specific environment that caused the film to form in the first place, it usually does not retain its passivity.

Oxygen may also play a diametrically opposite role and *increase* corrosion. This happens when oxygen is available at the cathode where it is reduced, thereby using up electrons. This keeps the corrosion reaction briskly progressing.

Metal and alloy classifications. Metals and alloys can be divided into two categories, as follows:

1. *Active-passive metals and alloys.* Metals that have the ability to become passive are termed *active-passive* metals. Steel, aluminum, chromium, stainless steels, nickel, nickel-base alloys, and titanium alloys are included in this group. It should be noted that passivity depends on the environment as well as on the metal. A metal that is passive in one environment may not be passive in another environment. A passive metal may show corrosion rates in a given environment many times lower than the same metal in the active state. Sometimes, a very small change in one of the process variables can shift a metal from the passive to the active state. This is the reason why careful laboratory tests should be conducted, taking into account the range of process variables when the responses of an active-passive metal to a specific environment are not fully known.
2. *Nonpassive metals and alloys.* As the term indicates, nonpassive metals do not have the ability to become passive. Zinc, lead, copper, and copper-base alloys are examples of this category.

Effect of increasing solution oxidation. When oxidizing agents such as oxygen, ferric chloride, or cupric chloride ions are gradually added in increasing amounts to solutions contacting *nonpassivating* metals, the solution oxidizing power is thus gradually increased, and the corrosion rate will continue to increase linearly. [See Figure 1.4 (bottom).] This reaction explains why copper-base alloys and other nonpassivating alloys are susceptible to corrosion in solutions containing oxidizing agents.

On the other hand, if *active-passive* metals are treated similarly as the solution oxidizing power is increased, the corrosion rate will initially increase; however, an abrupt change then occurs, and the corrosion rate decreases rapidly (by as much as 1000 times), even though more and more oxidizer is being added. [See Figure 1.4 (top).] The corrosion rate then remains low and constant as the oxidizers continue to be increased. Finally, at a very high concentration of oxidizer, the corrosion

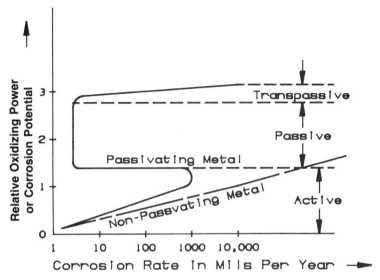

FIGURE 1.4 — *The effect of increasing solution oxidation (———) on passivating and (— — — —) nonpassivating metals.*

rate starts to increase again. [Note the active, passive, and transpassive zones in Figure 1.4 (top).] It should be noted that the proper amount of oxidizer must be used to keep the metal in the passive state. This is the basis for anodic protection where the amount of oxidizing is controlled by increasing or decreasing an applied potential on the surface to be protected (the anodic surface). (See Chapter 6.4.) The stainless steels have wide application in industry because of their active-passive character. They have excellent resistance to aerated acids and acids containing oxidizers, such as ferric and cupric ions. Stainless steels have excellent resistance to nitric acid, an oxidizing acid. The active-passive nature also explains some of the weaknesses of stainless steel. For instance, boiling fuming nitric acid, which is a very potent oxidizer (pushing the stainless steel into the transpassive region), will attack stainless steel. The active-passive metals and alloys will also be attacked in air-free solution. (Not enough oxidizers to push these materials out of the active zone.)

Another drawback of the active-passive metals is the formation of active and passive areas close together on the surface of the metal. These differences, as discussed earlier, create anodic and cathodic areas; thus, aggressive local corrosion occurs. For these reasons, when designing equipment of stainless steel, titanium, or other active-passive metals, particular care must be exercised to avoid designed-in crevices. In addition, it is necessary to design equipment that will prevent the accumulation of dirt, debris, etc., which can break down the protective film and produce active areas.

☐ 1.5 Frequency of Corrosion Failures

The frequency of corrosion failures attributed to the various forms of corrosion has been studied by the E. I. du Pont de Nemours & Co., Inc., Materials Engineering Group[10] for seven years based on reports from 23 company materials engineers. A report by Collins and Monack describes the results for the first four years of this study.[10] Table 1.3 shows an update on this study through the seventh year. This study indicates frequency of corrosion failures in one company in the chemical industry. Other companies and industries may show very different frequencies.

Failure Percentages

The first three forms of corrosion failures, namely general corrosion, stress corrosion cracking, and pitting, constitute 65% of the over 1200 cases reported. The frequency of failure from general corrosion (31%) and pitting (10%) is probably indicative of most industries in general and is not limited to the chemical industry alone. However, the frequency of stress corrosion cracking (24%) is probably excessive in relation to other industries, because the chemical industry relies heavily on austenitic stainless steels as a material of construction for many highly corrosive applications. Thus, where there is a preponderance of these steels, one might expect a higher frequency of this kind of corrosion. The next four types of corrosion failures, namely intergranular, erosion-corrosion, weld corrosion, and temperature, constitute 24% of the reported cases. The last seven kinds of corrosion failures make up only 11% of the cases.

Dealing with General Corrosion

General corrosion is defined as corrosion that attacks the surface of metals evenly and uniformly. According to the Du Pont survey, it is the most prevalent form of corrosion. The designer should deal with this

---

[10]E. I. du Pont de Nemours & Co., Inc., Engineering Department, Engineering Service Division, Wilmington, Delaware.

TABLE 1.3
Metal Failure Frequency for Various Forms of Corrosion

Forms of Corrosion Failure	Occurrences (%)	
General	31	
Stress Corrosion Cracking	24	65%
Pitting	10	
Intergranular Corrosion	8	
Erosion-Corrosion	7	24%
Weld Corrosion	5	
Temperature (cold wall, high temperature, and hot wall)	4	
Corrosion Fatigue	2	
Hydrogen (embrittlement, grooving, blistering, and attack)	2	
Crevices	2	
Galvanic	2	
Dealloying or Parting	1	11%
End Grain Attack	1	
Fretting	1	
TOTAL	100	100%

type of corrosion by selecting optimum materials of construction through adequate testing and then assigning a *corrosion allowance* to the equipment that is being designed. This subject will be covered in greater detail in Chapter 2.3.

Dealing with Other Corrosion Forms

All kinds of failures listed can be best controlled by the designer by understanding the conditions under which they occur and designing to avoid such conditions. Chapters 3 and 4 of this book will discuss in detail the various kinds of corrosion failures and what the designer can do to prevent them. Although certain corrosion forms showed a low-failure frequency rate in the Du Pont study, this does not mean that the designer should not consider them. For example, there could be 100 kinds of general corrosion failures that gradually leaked and were repaired one at a time without any plant shutdowns resulting, compared to, for example, one catastrophic hydrogen embrittlement failure that could suddenly shut down an entire plant.

■ 2 — SELECTION OF MATERIALS OF CONSTRUCTION

Selection of the optimum material of construction for a specific application is critical to the designer from the performance, safety, and economic standpoints. Conducting tests in a laboratory or pilot plant, relying on published corrosion rates from technical sources or producers' literature, or basing selection on previous company experience are ways in which optimum materials are selected.

☐ 2.1 Selection by Laboratory or Pilot Plant Testing

When no company experience exists, corrosion tests are needed to aid in the selection of materials of construction for corrosive service in plants or processes. The specific tests to be conducted from among the various corrosion tests available will depend, of course, on the expected process conditions. Preliminary testing is generally conducted in the laboratory using several candidate materials. Representative material samples are fabricated into several types of corrosion specimens. The specimen type is based on the form of attack expected, such as general, pitting, intergranular, crevice corrosion, end grain attack, weld decay, galvanic corrosion, stress corrosion cracking, dealloying, and cavitation or impingement attack. Of course, when the kind of attack is unknown, several specimen types are usually included in the preliminary tests.

When a pilot plant for a specific process is available, the testing can be more accurate. A selection of several candidate metals and alloys will be made based on the laboratory tests, a study of process flow sheets, and the materials' experiences. The same corrosion specimens are used.

☐ 2.2 Preparation and Assembly of Corrosion Test Specimens[11]

Figure 2.1 shows the following three steps required for preparing a standard rectangular specimen: (1) rough polish, (2) drill and stamp, and (3) final fine polish. Preparing welded specimens, which is also shown in Figure 2.1, requires a full penetration weld to be located on the center line of the specimen and finished in the same manner. Figure 2.2

shows stress corrosion or horseshoe specimens, two examples of specimens showing crevices and two dissimilar materials secured together to simulate galvanic corrosion. Figure 2.3 displays specimens designed for heat exchange surfaces, i.e., trial tubes in a mock-up heat exchanger, specimens clipped over heating coils, and the hairpin and bayonet types of heater specimens.

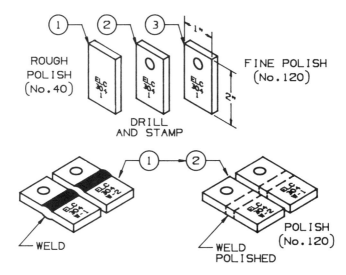

FIGURE 2.1 — *Preparation of standard test specimens for corrosion testing.*

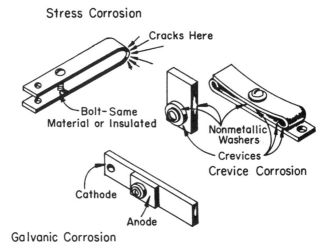

FIGURE 2.2 — *Specimens for stress, crevice, and galvanic corrosion.*

Various specimens may be assembled into racks for easily accessible equipment, as shown in Figure 2.4. The spool piece depicted is designed to insulate the specimens from each other to avoid possible galvanic corrosion. The simplest procedure involves tying the specimens together with a resistant nonconducting material, such as a polytetrafluoroethylene (PTFE) cord, and suspending them in the process solutions. The bracket assembly is designed to be attached to rods or pipes. Figure 2.5 shows two kinds of specimen racks for limited access that are typical of blank-nozzle and threaded-plug-insert designs. Figure 2.6 displays typical specimen racks for pipelines.

Laboratory Procedure

For any of the specimens mentioned above, general laboratory procedure is as follows:

1. Measure the area and clean, dry, and weigh specimens when required.
2. Observe specimen surfaces. Frequently, color photographs are taken of critical specimen surfaces, such as end grain, tension side of stressed specimens, and welds. This avoids confusion when appraising the results on the specimens. For example, specimen surfaces before testing often contain pits.
3. Then conduct the test in the test solution, simulating the service environment as closely as possible.
4. Periodically observe the specimens, particularly stressed specimens, and in some cases, intermediate weighings are performed. Checking the specimens during the test is important because corrosion rates often vary with time.
5. When the test is completed, carefully clean, dry, observe, and (when required) weigh the specimens.
6. Calculate corrosion penetration rates and draw conclusions based on the observations made.
7. Select candidate materials.

Material Selection in the Pilot Plant

The pilot plant is an excellent place in which to choose the best of several materials that, because of the laboratory tests, had been selected previously. The pilot plant's process conditions simulate the actual conditions more closely than laboratory tests. For example, suppose that in a specific chemical process, three solutions, Solutions A through C, are added one at a time at three different temperature

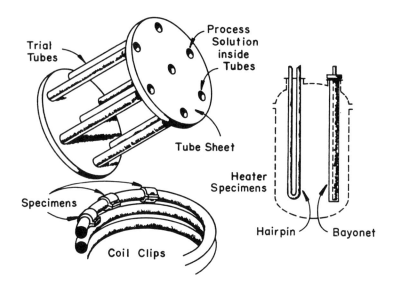

FIGURE 2.3 — *Specimens for heat-exchange surfaces.*

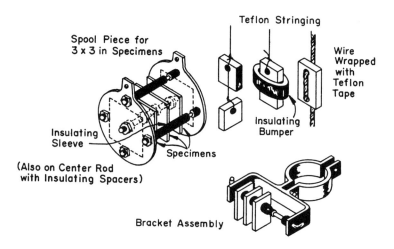

FIGURE 2.4 — *Specimen racks for easily accessible equipment.*

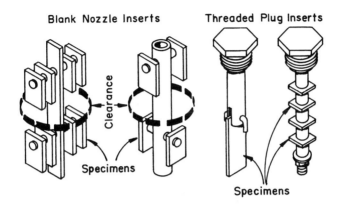

FIGURE 2.5 —*Specimen racks for limited access.*

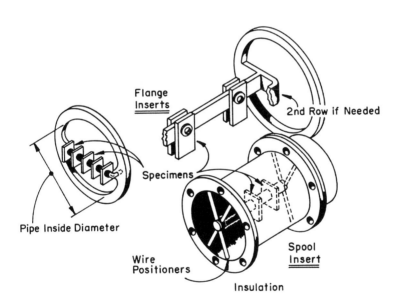

FIGURE 2.6 — *Specimen racks for pipelines.*

levels over a period of time to a reactor for which the material of construction is to be selected. For thorough laboratory testing, the following tests might be contemplated. Candidate specimens can be exposed to:

Test 1 — Solution A alone at the first temperature

Test 2 — Solutions A and B at the second temperature

Test 3 — Solutions A through C at the third temperature

Such an approach would entail a multiplicity of individual tests so a compromise might be made to save time and costs. For example, one test might be conducted with Solutions A through C at the highest reactor temperature expected. Such a test would not closely simulate process conditions.

On the other hand, the pilot plant is a miniature producing unit that can expose the corrosion test specimens to the actual plant's process conditions. Specimens of the same corrosion type, as discussed previously, can be installed throughout the pilot plant.

Special Tests

Materials shown to be acceptable by the pilot plant corrosion tests may not always be suitable for final plant use. When velocity, temperature, and hot-wall effects expected in the anticipated process are difficult or impractical to duplicate in regular corrosion tests or in the pilot plant, the following special testing equipment is available:

Velocity tester. Although it is generally agreed that the velocity of a corrosive stream can have a significant effect on corrosion rates, there is a lack of specific corrosion data in the technical literature on this subject. Extrapolation of rates from lower velocity data to higher velocities can be meaningless. Rule-of-thumb methods are sometimes used to counteract this problem. For example, it is common practice in replacing carbon steel pipelines that have failed in a short time in 60° Baume (78%) sulfuric acid service to increase the pipe size by 1/2 to 1 in. (13 to 25 mm). This change may mean the difference between a few months and many years of service life. To satisfy the need for a reliable velocity test and to avoid the inherent drawbacks of other test methods, a simple dynamic corrosion test was developed at the Du Pont Engineering Test Center (Newark, Delaware).[12] The specimen holder shown in Figure 2.7 is a Teflon[†] polytetrafluoroethelyne (PTFE) cylinder

---

[†]Registered trade name.

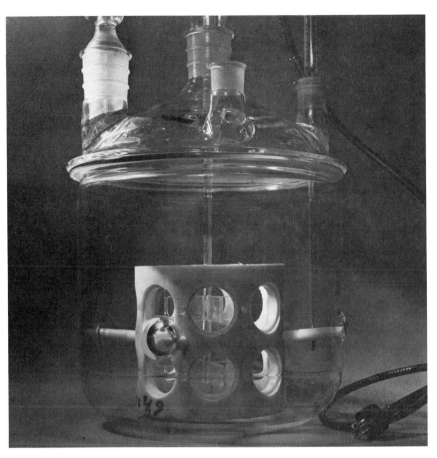

FIGURE 2.7. *Velocity test apparatus.*

with 12 round openings that receive Teflon mushrooms in which the circular specimens are mounted. Only the specimen faces are exposed. Teflon bolts are screwed into the Teflon cylinder. They fit into four depressions in the wall of the test jar and restrain the cylinder from rotating. Figure 2.8 exhibits another single tier test unit in operation. A specially designed glass agitator, powered by a DC variable speed motor, causes the solution to flow across the specimen faces at a predetermined velocity. Heat is supplied by a heating mantle shown under the test jar. In front of the unit, a specimen in a mushroom holder is shown. Other specimens in their individual holders are installed in a large Teflon ring. Details of the construction of these velocity test units are outlined in NACE Standard TM0270-70.10 ("Method of Conducting Controlled Velocity Laboratory Corrosion Tests").

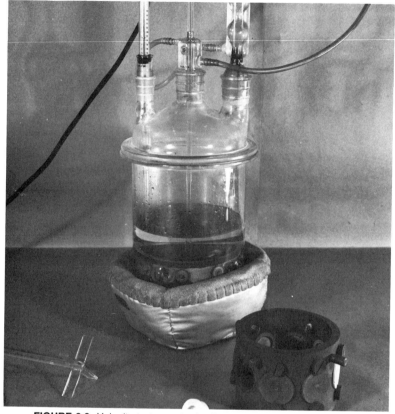

FIGURE 2.8. *Velocity tester; note the "mushroom" specimen holder and glass agitator located in front of the unit.*

High-temperature and high-pressure tests. Another type of special test simulates conditions of high temperature and high pressure. Frequently, to increase yields, the temperature of a process is increased. Serious corrosion attack may result if the effect on the materials of construction is not considered. A good example is the effect temperature has on the corrosion resistance of austenitic stainless steels in 62% nitric acid. At temperatures below the boiling point, resistance is satisfactory, but at higher temperatures, stainless steel is rapidly attacked. At 375 F (191 C), for instance, the corrosion resistance of stainless steel is very poor; however, titanium performs well at that temperature.

Figure 2.9 shows a simple test unit for conducting tests up to 450 F (232 C) and 600 psig that was developed at Du Pont's Engineering Research Laboratory (Wilmington, Delaware). This equipment is made from a 2-in.- (51-mm)-diameter pipe flanged at both ends and a 60-mil

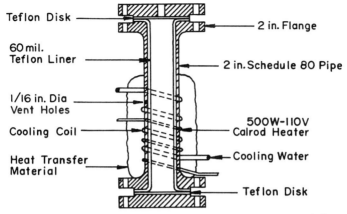

FIGURE 2.9 — *Teflon-lined vessel for corrosion tests in hot solutions.*

Teflon sleeve located inside the test chamber. The sleeve is formed over the flange faces making a Van Stone joint when the bottom and top blind flanges are secured into position. The blind flanges are also protected by Teflon discs. The top flange is drilled to expose the Teflon sheet, which acts as a safety disc for excessive pressures. Heat is supplied by a 500-ampere, 100-volt Calrod[†] heater wound around the unit. A thermocouple on the pipe wall controls the temperature. A covering of heat transfer material completes the unit.

This type of unit was used for many years with good results. An up-to-date version of a high-temperature, high-pressure reactor used for corrosion testing is displayed in Figure 2.10. This unit can be used at pressures of up to 1900 psig (13 MPa) and temperatures up to 350 C. The test chamber can be made of any of nine different alloys. A modern controller accurately moderates the temperature.

Hot-wall tester. Many failures have occurred because the effect of a heating coil's hotter wall in a corrosive environment was not considered. Circular specimens that clip around a pipe being heated (as shown in Figure 2.3) are an approach to this test problem, but are not entirely satisfactory because of the unavoidable space between the specimen and the pipe. This space reduces heat transfer and hence produces a lower specimen temperature than what will occur in actual practice. The other specimen types, also shown in Figure 2.3, are satisfactory, but are somewhat more expensive because such tests are time consuming. A very simple test setup designed to simulate closely

---

[†]Registered trade name.

FIGURE 2.10 — *A modern version of a high-temperature, high-pressure reactor used for corrosion testing.*

the hot-wall effect has been used by many laboratories.[13] The specimen is heated directly by a heating iron and is mounted and sealed in the bottom of a glass vessel that contains the solution to be heated. The specimen acts as the heat transfer surface and maintains a true "hot-wall" temperature as it heats the solution to the required temperature level. This test is very economical to conduct and produces rapid results.

Former Plant Experience

Many times, materials of construction for a new plant are selected solely on the basis of a similar plant producing the same product. Such experience is usually better than relying on corrosion tests or pilot-plant-generated corrosion data. However, the designer should decide whether or not conditions are actually identical. The designer should ask the following questions:

- Is the water supply the same?
- Will the incoming materials be the same? Will they be of the same purity if a different supplier is used?
- Are soil conditions the same (for buried pipe)?
- Is the atmosphere the same? Will the new plant be closer to the sea? Will the new neighbor's pollution be different?
- Will electrical generation units that can cause stray currents be close by?

(Later sections of this book will discuss each of these questions.)

These and other questions should be asked before the final selection of materials of construction is made. The following example depicts what can happen if these questions are not considered:

> *Example* — A western plant was built in which all the materials of construction were identical to those of another plant in the East producing the same product. Shortly after the plant began using the materials, the carbon steel pipelines, heat exchange tubing, and other carbon steel equipment handling the plant water failed disastrously because of corrosion. The available water source was determined to be extremely aggressive. The plant was shut down, and the carbon steel had to be replaced by stainless steel. The considerable expense of this mistake could have been avoided if someone had questioned whether or not the water supplies of the two plants were the same.

Published Corrosion Rates

Because of a lack of time or testing facilities, selection of materials of construction is sometimes based solely on published corrosion rates. A problem can exist with this approach if the test details are inconsistent with the expected process conditions and the tests are not conducted by a competent laboratory. Much of the published corrosion rate data are from beaker immersion tests. If the environment involved is quiescent, the corrosion rates may apply, but if fluid velocity will be encountered, the corrosion rates may be very unreliable. Very seldom does the literature report the character of attack, which is very important. For instance, suppose that the corrosion specimens are attacked by pitting. Corrosion rates based on these specimens would be very low and misleading and therefore hazardous to use in material selection.

If the published corrosion rates were determined using relatively pure solutions (which is usually the case), they would not reflect corrosion rates in similar streams which, among other things, contained trace impurities and suspended solids, and were aerated differently than the test solution.

Trace impurities. Trace impurities which may be in plant process fluid streams but not in the solutions where the corrosion rates were determined can increase or decrease the actual corrosion rates. For example, small amounts of copper in solution can deposit as metallic copper on aluminum and steel surfaces and accelerate corrosion. Austenitic stainless steels, on the other hand, are inhibited by copper if it is in the oxidized state. Ferric ions can also accomplish this for stainless steel. Depending on whether a particular material functions best in oxidizing or reducing conditions, the presence of even small amounts of oxidizing or reducing agents can greatly affect the materials' corrosion performance, as discussed in Chapter 1.

Suspended solids. The presence of suspended solids may increase corrosion attack in piping, valves, and pumps which would ordinarily show low rates of attack in published corrosion results of the materials tested in cleaner solutions.

Aeration. The effect of aeration often promotes rapid attack in some materials and lowers attack in others, as was also discussed in Chapter 1. Therefore, whether or not you expect aeration in your process should be determined before relying on published data. The published corrosion rates used for materials determination should state the precise conditions under which the tests were conducted. (This is seldom done.)

Solution pH. The acidity of a solution can have a substantial effect on the aggressiveness of the solution. The hydrogen ion concentration of a solution is designated as pH and is defined as:

$$pH = -\log[H^+] \tag{2.1}$$

The lower the pH, the greater the hydrogen ion concentration and *vice versa*. A pH of 7 is termed *neutral* with alkaline conditions above that figure and acid conditions below 7. Often in industry, where a process solution is low in pH, alkalis are added to the solution to raise the pH and, hence, in many applications, to lower the corrosivity of the solution.

For some metals and alloys, a small shift in the pH of a solution can be the difference between low and high corrosion conditions. Examples include steel, aluminum (which is amphoteric and can be attacked by both alkalis and acids), and magnesium. Conversely, examples of metals resistant to a wide range of pH conditions are tantalum and titanium.

Pourbaix diagrams are available for a number of metals and alloys which indicate, based on thermodynamic equilibria, whether or not corrosion can be expected under different pH-potential conditions.

The best advice is that if the process streams are expected to be aggressive, taking sufficient time to conduct corrosion tests yourself rather than relying on published corrosion rates will be less expensive in the long run. Even if your organization does not have the necessary facilities, a commercial laboratory can conduct corrosion tests for you.[14] Publishing corrosion rate data performs a valuable service because it facilitates the selection of candidate materials to test.

Nelson Charts

Before the advent of the Nelson corrosion charts, interpreting published corrosion data was very cumbersome and confusing. For instance, the following is a bulky and confusing description of how aluminum reacts to nitric acid:

Aluminum corrodes rapidly in nitric acid, unless the concentration at room temperature is below 0.5%; then the corrosion rate is below 20 mils per year (mpy). From 0.5 to 80%, corrosion is very high. However, above 90%, corrosion decreases to 20 mpy. As the concentration rises to about 99.5%, the corrosion rate decreases to less than 2 mpy. However, if the temperature elevates to 125 F (52 C), the rate increases to 50 mpy if the concentration is between 90 and 100% acid. At temperatures of about 175 F (79 C), corrosion rates are again very high.

George A. Nelson devised simple corrosion rate charts that show this kind of corrosion information at a glance. Industry is indebted to him for his valuable contribution. Figure 2.11 shows the key to using a Nelson chart. Figure 2.12 is an actual Nelson chart, which shows how aluminum reacts to nitric acid. Note how simple and understandable it is compared to the description written above. Figure 2.13 is a page from *Corrosion Data Survey — Metals Section*, a book which identifies metals that may have satisfactory corrosion performance in given environments using the Nelson technique of recording corrosion data. This book shows various corrosion rates of iron-base, copper-base, nickel-base, and other metal alloys in many environments. It is available from the National Association of Corrosion Engineers (NACE)[(1)] and is a very valuable source of corrosion information. Every design office should have a copy. The survey is also available in computerized form from NACE (COR·SUR^{TM}).

[(1)]All organization names and complete addresses are listed in the Appendix.

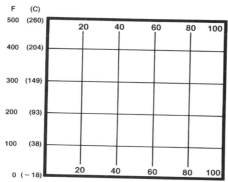

Matrix Key

Data points representing average penetration per year (key below) are plotted on the matrix enlarged here. The horizontal grid represents percent concentration in water and the vertical grid represents temperature.

Average Penetration Rate Per Year

Code		Mils	Inches	μm
●	<	2	0.002	50
○	<	20	0.020	508
□	{	20-50	0.020-0.050	508-1270
X	>	50	0.050	1270

Key to Data Points

FIGURE 2.11 — *Key for using Nelson charts.*

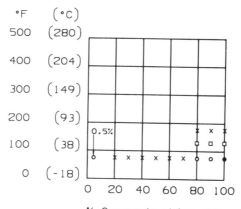

FIGURE 2.12 — *Nelson chart showing how nitric acid affects aluminum.*

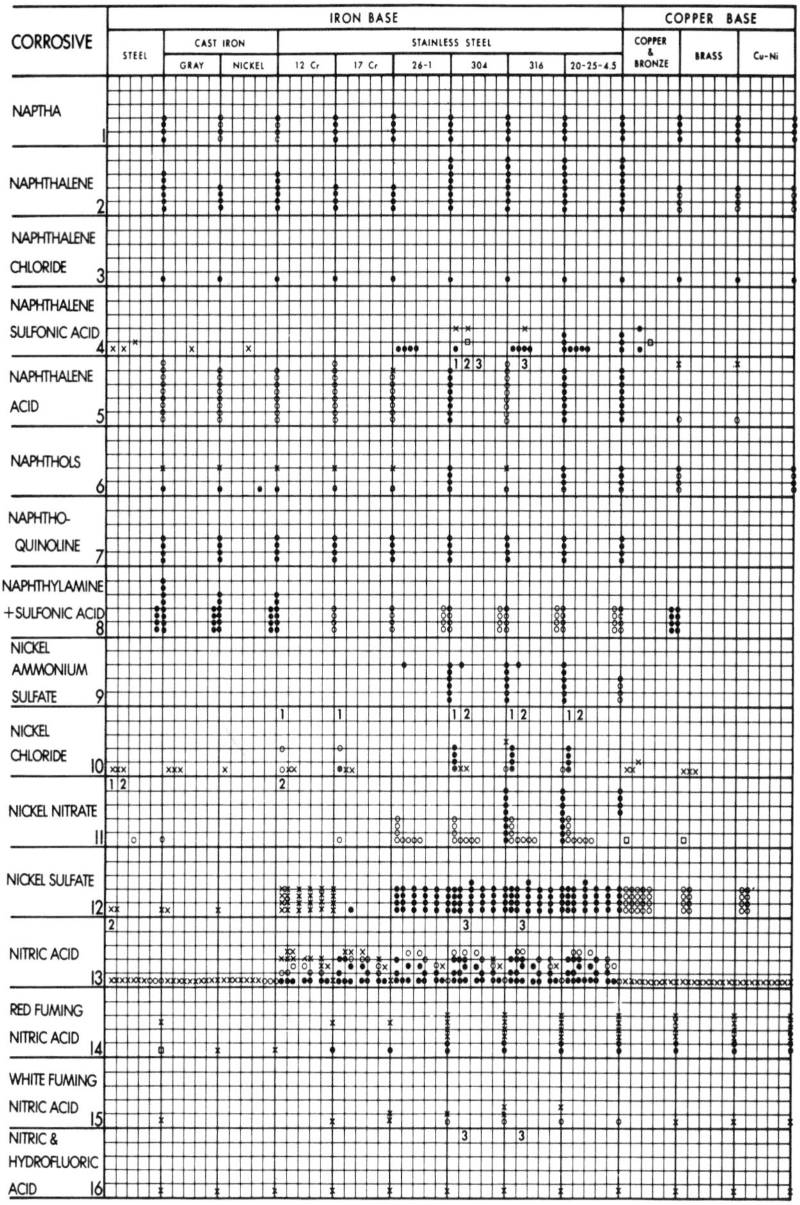

FIGURE 2.13 — *Page from Corrosion Data Survey.*

Construction Materials Producer — A Corrosion Information Source

The producers of construction materials are another good source of corrosion information. Many of these producers have conducted extensive corrosion tests to determine the corrosion rates of their products in various environments. Much of this kind of material information and many application recommendations are available upon request. Table 2.1 lists producers of various materials of construction which may furnish corrosion information on their products.

☐ 2.3 Corrosion Rate Determinations and Allowances

Corrosion Rate Derivation and Calculation

Ever since man became interested in the corrosion of metals, there has been a need for a convenient way of designating just how much or how little a material has corroded. Older texts on corrosion referred to "loss of milligrams per square decimeter per day." These designations were not very useful except for comparison purposes because it was difficult to visualize what the effect would be on process equipment. Today, the expected corrosion penetration into the wall of a vessel or tubing caused by corrosion is usually recorded as inches penetration per year or mils [0.001 in. (0.03 mm)] penetration per year. This corrosion rate designation is used throughout industry. By knowing the weight loss of a corrosion specimen over a number of days and the overall area and density of the specimen, the designer can determine a penetration into the specimen surfaces by corrosion. The derivation of and formula for corrosion penetration rates are listed in Table 2.2. Today, much corrosion rate determination is done through computers.

Corrosion Rate Nomographs

A nomograph[15] used for corrosion rate calculation is exhibited in Figure 2.14. Although considered obsolete, a nomograph provides an easy means for calculating corrosion rates based on the formula noted in Table 2.2. To use the nomograph, first locate the corrosion loss in grams per square inch on Axis A. Then lay a straight edge x on this number and the duration of testing days on Axis C. Where x crosses Axis B, lay straight edge y between that point and the point on Axis E which is the density of the material involved. Read the corrosion rate in inches penetration per year where y crosses Axis D.

TABLE 2.1 — Corrosion Information Sources for Metals and Alloys

Stainless Steels

Climax Molybdenum
P.O. Box 1568
Ann Arbor, MI 48106
(Ask for M-11 Guide to
Corrosion Resistance)

Eastern Stainless Steel Company
Rolling Mill Rd.
P.O. Box 1975
Baltimore, MD 21203

Jessop Steel Co.
Jessop Place
Washington, PA 15301

Armco, Inc.
703 Curtis
Box 600
Middletown, OH 45043

Teledyne Vasco
Box 151
Latrobe, PA 15650

Bethlehem Steel Corp.
Bethlehem, PA 18016

Carpenter Technology Corp.
Carpenter Steel Division
101 W. Bern St.
Reading, PA 19603

Nickel and Nickel Alloys

Inco Alloys International
Guyan River Rd.
Huntington, WV 25720

Cabot Wrought Products Division
Cabot Corp.
1020 W. Park Ave.
Kokomo, IN 46901-9013

Teledyne Allvac
2020 Ashcraft Ave.
Monroe, NC 28110

International Nickel Co.
Park 80 West Plaza Two
Saddlebrook, NJ 07662

Aluminum

Aluminum Company of America
1501 Alcoa Bldg.
Pittsburgh, PA 15219

General Extrusions, Inc.
P.O. Box 2669-E12
Youngstown, OH 44707

Alcan Aluminum Corp.
100 Erieview Plaza
Cleveland, OH 44114

Kaiser Aluminum and Chemical Co.
300 Lakeside Dr.
Oakland, CA 94643

Reynolds Metals Co.
P.O. Box 27003
Richmond, VA 23261

Aluminum Association
900 19th St., N.W. Suite 300
Washington, DC 20006

Carbon Steel

Bethlehem Steel Corp.
Bethlehem, PA 18016

United States Steel Corp.
600 Grant St.
Pittsburgh, PA 15230

Armco, Inc.
703 Curtis
Box 600
Middletown, OH 45043

Latrobe Steel Co.
2626 Ligonier St.
Latrobe, PA 15650

Inland Steel Co.
30 W. Monroe
Chicago, IL 60603

Copper and Copper-Base Alloys

Amax Copper, Inc.
200 Park Ave., 51st Floor
New York, NY 10017

Mueller Brass Co.
1925 Lapier Ave.
Port Huron, MI 48060

Ampco Metal
An Ampco-Pittsburgh Co.
Dept. 1702
P.O. Box 2004
Milwaukee, WI 53201

Revere Copper Products Co.
P.O. Box 300
Rome, NY 13440

Abex Bronze and Alloy Co.
P.O. Box 458
Meadville, PA 16335

Copper Development Association
2 Greenwich Office Park, Box 1840
Greenwich, CT 06836

Titanium and Titanium Alloys

Titanium Metals Corp.
Timet Division
400 Rouser Rd.
Pittsburgh, PA 15230

Oremet Titanium
530 W. 34th St.
P.O. Box 580-ME
Albany, OR 97321

Industrial Titanium Corp.
3041 Commercial Ave.
Northbrook, IL 60062

RMI Titanium
1000 Warren Ave.
Niles, OH 44446

Teledyne Wah Chang
P.O. Box 460
Albany, OR 97321

Kennametal, Inc.
P.O. Box 346
Latrobe, PA 15650

TABLE 2.2
Corrosion Penetration Rate

Derivation:

$$\text{Penetration Rate} = \frac{\text{Weight Loss (W)}}{\text{Area (L}^2) \times \text{Density } \frac{W}{L^3} \times \text{Time}}$$

$$= \frac{W}{L^2 \times \frac{W}{L^3} \times \text{Time}}$$

$$= \frac{W \times L^3}{L^2 \times W \times \text{Time}} \quad \text{(canceling out)}$$

$$= \frac{L}{\text{Time}}$$

Calculation:

Mils Penetration per Year (mpy) = $\frac{534 \times W \times 1000}{A \times d \times t}$

Where: W = weight loss in grams
 A = total surface area in square inches
 d = density of specimen in grams per cubic centimeter
 t = time of exposure in hours
 L = length or dimension used for area or density.

NOTE: Where area is determined in square centimeters, multiply by 6.45.

Average Densities g/cm^3:

Aluminum (Type 3003)	2.74	Cupro-Nickel (70:30)	8.94
Red Brass (85%)	8.75	Nickel	8.90
Silicon Bronze	8.53	AISI 304 Stainless	7.93
Cast Iron Gray	7.19	AISI 430 Stainless	7.70
Copper	8.96	Low Carbon Steel	7.86
		Titanium	4.54

The nomograph can also be used for converting grams per square decimeter per day (g/dm^2/day) to inches per year (in./y).

Corrosion Allowance

By knowing the expected general corrosion rate and the anticipated plant life, the designer can calculate the extra wall thickness required for corrosion resistance of the process equipment he is designing. After

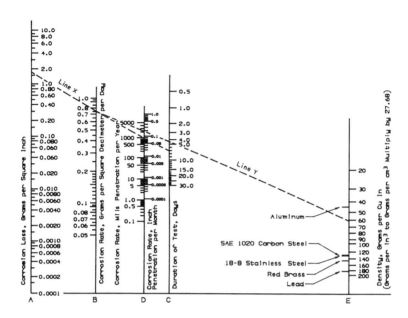

FIGURE 2.14 — *Nomograph for calculation of corrosion rates. (From Reference 15.)*

determining a wall thickness that meets mechanical requirements, such as pressure and weight of equipment, an extra thickness called a *corrosion allowance* is added to the wall thickness to compensate for the metal expected to be lost over the equipment's life. Then, because the penetration depth may vary, a corrosion allowance is assigned a safety factor of 2. CAN BE OFF (2)

As an example, suppose a tank wall required a 3/16-in. (5-mm) wall thickness for mechanical considerations. The designer has determined that the corrosion rate will be 15 mpy and the expected life of the equipment will be 10 years. The total corrosion allowance is 0.015 in. (0.4 mm) (corrosion rate per year) x 10 (years) = 0.15 in. (4 mm). The corrosion allowance is doubled to 0.3 in. (8 mm) as a safety consideration. The final wall thickness would be 0.3 + 0.1875 = 0.4875 in. (8 + 5 = 13 mm). The designer would then specify a 1/2-in. (13-mm) wall thickness as the closest standard plate available.

Estimated Plant Life

Most plants are designed to function for 10, 15, or, in some cases, 20 years or more. Corrosion allowances, even taking into account some non-uniform penetration, have proven to be very useful in the design of process equipment.

However, it must be emphasized that the use of corrosion allowances in predicting plant life are only meaningful if general corrosion is imminent where substantially uniform corrosion will occur. Based on the Du Pont study result (30% general corrosion, 70% other corrosion), it is predicted that 70% of corrosion incidents in industry will be of a more local nature, such as stress corrosion cracking, pitting, intergranular corrosion, crevice corrosion, dealloying, etc.

☐ 2.4 Corrosion Rate Classification

When a designer is selecting a material of construction for use in a specific process solution and he has the corrosion rate figures for a number of candidate alloys, how does he determine whether or not the materials have adequate resistance? How does he determine whether a less expensive material can be used instead of the more expensive one? Table 2.3, which classifies corrosion rates, has been compiled to help answer these questions. This table has been successfully used as a guide for many years.

When to Use Metals and Alloys with Certain Corrosion Rates

Different metals and alloys have varying corrosion rates depending, of course, on specific environments. This will be discussed further in later chapters. The following guidelines (following corrosion rates established in Table 2.3) list the maximum suggested corrosion rates for specific working environments.

1 mpy maximum. Use 1 mpy maximum when product contamination can be a problem. This is a very low corrosion rate. The manufacture of fine silica is a good example for this classification. The product must be snow white, even 10 parts per million (ppm) of iron will discolor the product. Stainless steels are used as well as other corrosion-resistant materials, because even though carbon steel would have resisted the environment with a corrosion rate of 12 mpy, the amount of corrosion would be enough to discolor the product.

Food industry equipment also allows very little contamination. Therefore, almost completely resistant metals are used. An added problem is that frequent cleaning using live steam and cleaning solutions is required. Obviously, the material must resist the cleaning

TABLE 2.3
*Characteristics and Uses
of Corrosion Rates*

Rate in Mils Penetration Per Year (mpy)	Characteristics and Uses
1 maximum	Very low corrosion; adhered to for services where product contamination is a problem.
10 maximum	Low corrosion; adhered to for thin-walled process equipment.
20 maximum	Fairly low corrosion; can be considered the "normal" maximum allowed in chemical equipment.
50 maximum	High corrosion; seldom tolerated except in thick-walled equipment where product contamination is controlled.
Rates over 50	Excessive corrosion; very seldom tolerated and only then in very thick-walled equipment where massive product contamination is not a problem.

process, which may be much more aggressive than the process conditions.

Where very expensive materials like tantalum and platinum are used, corrosion rates should be practically zero. Parts with critical dimensions, such as valve seats, orifices, and spray nozzles, should also be limited to a 1 mpy maximum corrosion allowance.

10 mpy maximum. Use 10 mpy maximum where thin-walled equipment is to be used. This rate is low but is required for certain diameters of Schedule 5S pipe and tubing (such as heat exchanger tubing) and other extra-thin equipment. Today, much Schedule 5S stainless steel pipe is used to reduce costs.

20 mpy maximum. Use 20 mpy maximum for general applications. Although this rate is considered to indicate fairly low corrosion, it is the normal maximum in the chemical industry. The majority of industrial applications will probably be in this category.

50 mpy maximum. Use of 50 mpy maximum is very infrequent and then only for thick-walled equipment where product contamination is either not a problem or can be adequately controlled. This corrosion rate indicates a high amount of corrosion.

Rates above 50 mpy maximum. Corrosion rates above 50 mpy should not be considered unless definite evidence for an economical use exists. There have been very few exceptions to this rule. An example of an exception is cast iron used in the production of muriatic acid. Sulfuric acid is dripped onto a bed of sodium chloride in a unit called a *Mannheim furnace* operating at a high temperature. Rotating ploughs are used to keep the salt moving and mixing so that the chemical reaction will be uniform and thorough. The rotating parts and ploughs are made of cast iron. Although the corrosion rate of the cast iron is several hundred mpy, the ploughs have an adequate service life because the cross sections are greater than 3 in. (76 mm). Cast iron is the most economical material to use, despite its high corrosion rate in this application. It should be added that product contamination is not a problem in this process because the corrosion products are carried away by the remaining "soup" of sodium sulfate, while the uncontaminated hydrogen chloride gas goes overhead. This process is now obsolete.

3 — DESIGN SOLUTIONS TO CORROSION PROBLEMS BASED ON TYPE OF CORROSION

☐ 3.1 Crevice Corrosion

Corrosion in crevices is an insidious form of deteriorization which can be controlled by careful design and fabrication. This type of corrosion is actually a particular form of concentration cell corrosion. Because it is very common, crevice corrosion is generally considered a class of corrosion by itself.

A crevice may, of course, be formed in many ways, such as by a metal contacting another metal or a nonmetal such as a gasket or simply an inadvertant crack in a metal. Most metals and alloys, with the exception of a few noble ones, are susceptible to this form of attack. Figures 3.1 and 3.2 show inside and outside views of how a crevice in a cast iron pipe caused corrosion from the outside into the interior of the pipe.

Crevice Corrosion Mechanism

A crevice can cause localized corrosion by providing a site for a process composition or a concentration inside the crevice, which is different from that outside the crevice. Thus, a corrosion cell is set up. Such differences cause a loss of metal at the anodic area. Corrosion caused by a difference in *oxygen concentration*, for instance, is shown in Figure 3.3(a). The solution inside the crevice, because it is isolated from the main solution, becomes low in oxygen. The solution outside the crevice is continually replenished with oxygen and therefore remains high in oxygen. An abrupt difference in concentration results, and a corrosion cell is formed. The anodic area is the area inside the crevice which corrodes. Figure 3.3(b) shows another type of crevice corrosion called a *metal ion concentration cell*. The metal ions concentrate in the crevice while the solution outside the crevice, because of solution velocity or process conditions, becomes much lower in metal ion concentration and again, a corrosion cell is formed. The area of the high metal ion concentration becomes the cathode, and the localized area

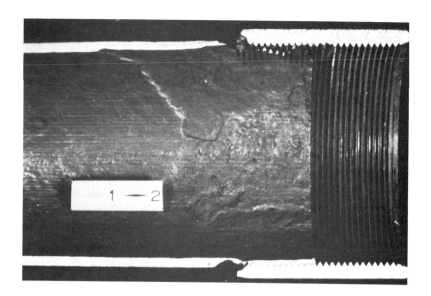

FIGURE 3.1 — *Inside view of corroded cast iron pipe; note that corrosion has progressed from the outside to the inside of the pipe.*

FIGURE 3.2 — *Outside view of the same cast iron pipe in Figure 3.1; example of crevice corrosion failure.*

outside the crevice forms the anode where corrosion occurs.

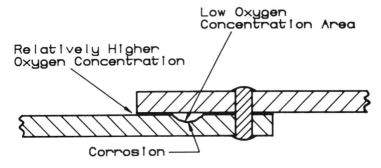

A. Oxygen Concentration Type

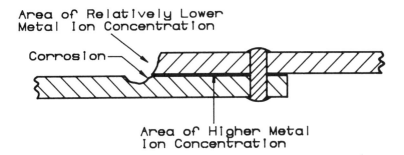

B. Metal Ion Concentration Cell Type

FIGURE 3.3 — *Two types of crevice corrosion; A. oxygen concentration and B. metal ion concentration cells.*

Certain metals and alloys are more susceptible to crevice corrosion than others. This form of corrosion occurs most commonly on metals that owe their corrosion resistance to protective films, such as stainless steels, titanium, and aluminum.

Welding and Potential Crevices

The fusion welding of metals and alloys can create crevices. For instance, when a single-butt weld is made, there is always a possibility that because of a lack of complete weld penetration, a crevice will form at the root of the weld. [See Figure 3.4(a).] When practical, the designer should specify double-butt welds to eliminate crevices, as shown in Figure 3.4(b). J-groove joints are usually as crevice free as butt joints; however, avoid socket joints because they leave an inherent crevice. Do not specify threaded joints for corrosive service because they also have an inherent crevice. If a lap joint is required, a single-lap joint should not be specified, as shown in Figure 3.4(c), but a double-lap joint should be required, as shown in Figure 3.4(d), to eliminate the crevice. In many pipelines, failures have occurred because of crevices left inside a pipe that had been welded with single-butt welds. Ideally, double-butt welds should be specified; however, it is not always possible to construct them. If the pipe diameter is large enough for a welder to get inside, double-butt welds can be made; however, most widely used pipe have much smaller diameters. Therefore, many pipelines are joined with single-butt welds.

Crevice corrosion caused by welding. A good example of what can happen if crevices are left after welding occurred at an Eastern chemical plant. The plant produced chlorosulfonic acid and shipped it in 55-gallon (208-liter), AISI[1] 302 stainless steel drums. The corrosion rate for the concentrated acid on this kind of stainless steel is practically zero. However, if the acid becomes diluted, it will attack stainless steel at a very high rate. A group of approximately 300 drums was received from a manufacturer. What was not noticed was that the single-butt girth welds used to attach the top and bottom heads to the bodies showed an almost complete lack of penetration at the tie-in area, as shown in Figure 3.5. Note the deep crevice in this area. Figure 3.6(a) shows the crevice depth at 5X magnification, which is more than halfway through the drum wall. Figure 3.6(b) shows a good weld at 5X magnification, The weld is located away from the tie-in area. This weld shows complete penetration and consequently no crevices. The drums were filled with acid and shipped to customers. After the acid was used up, the drums were dutifully washed out and shipped back to the producer. By the time the drums were returned to the producer, the welds at crevice areas had been vigorously corroded completely through the drum wall by the diluted acid that had been retained in the crevices from the washing operation. The drums consequently leaked badly at these areas when they were refilled. Most of these drums had to be scrapped. As a result of these crevice corrosion failures, the drum manufacturer improved his welding methods and techniques, and the buyer insisted on a more stringent inspection for other batches of drums. Figure 3.7 shows a

---

[1]American Iron and Steel Institute (AISI), Washington, DC; see Appendix for complete addresses of associations.

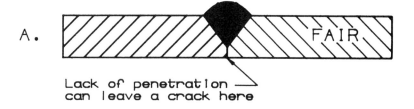

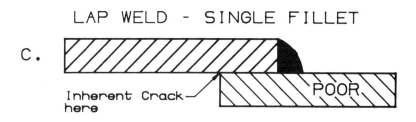
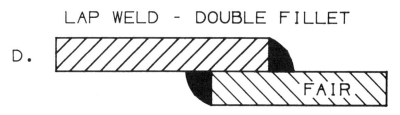

FIGURE 3.4 — *Types of weld joints: (A) single-butt weld, (B) double-butt weld, (C) lap weld — single fillet, and (D) lap weld — double fillet.*

bottom and top of a drum that was cut from the new batch of drums. Note that the welds are continuous and exhibit full penetration welds without crevices. No further crevice corrosion problems were experienced with the new drums.

FIGURE 3.5 — *Weld inside bottom.*

Use consumable insert rings. One alternative to aid the designer in ensuring that no crevices will form in single-butt welded pipe and tubing is to specify backing rings. However, the designer must be cautious about the type of ring he specifies. For instance, the standard backing ring may assure full penetration of the weld, but will leave crevices on both sides of the weld. [Refer to Figure 3.8(a).] These two crevices can cause as much trouble as a lack of weld penetration. For critical pipelines, consumable insert rings may be specified to assure full penetration welds with no crevices. One category of these rings called *EB inserts* was developed by the Electric Boat Company[2] to alleviate problems in piping systems of atomic powered submarines. Prior to the

---

[2] Electric Boat Company, Groton, Connecticut.

(a)

(b)

FIGURE 3.6 — (a) Tie-in area at 5X magnification showing poor penetration leaving a crevice; (b) weld at 5X away from the tie-in area showing excellent penetration and consequently, no crevice.

development of these rings, crevice corrosion of welded tubes plagued these power plants. [See Figure 3.8(b).] Another type of ring called *flat ring consumable inserts*, displayed in Figure 3.8(c), developed by the Grinnell Corporation,[3] has also been used with good results.

Another consumable insert ring called the *Y-insert* has proved efficient in eliminating crevices in pipelines. Figure 3.9(a) shows the insert tack welded into the weld groove ready for welding, and Figure 3.9(b) displays the results after the insert ring has been fused into place. This constitutes an initial weld pass, is flush with the inside diameter of the pipe, and leaves no crevices. The weld joint is then completed with several weld passes. This form of insert ring has been used with good results in eliminating crevices in buried pipelines.

All of these consumable insert rings are made from an alloy with

---

[3] Grinell Corporation, New Haven, Connecticut.

special metal compositions similar to that of welding filler rods used to weld specific alloys. When these rings are welded by competent welders using special techniques, complete fusion is attained and crevice-free welds result. (See American Welding Society's D10.4 "Recommended practices for Welding Austenitic Chromium-Nickel Stainless Steel Piping and Tubing" for more information on consumable inserts.)

Use removable backup rings. Another method of assuring full penetration welds is to specify backing rings that can be removed after welding. This kind of backup ring has proved practical in a number of instances. As an example, difficulty was experienced at a plant in welding a tube bank composing many return bends without leaving crevices at the welds. To complicate this further, a smooth bore was required. Welds without crevices were finally constructed using graphite backup rings. These rings were installed at each weld joint, and during welding, they allowed full penetration of welds as higher heats could be used without danger of burn-through. [See Figure 3.10.]

After welding, the rings had to be removed. Because of the pipe bends, the rings could not be simply pushed out; therefore, the problem was solved by a rather unique method. Before welding, a steel cable was threaded through the holes in the installed rings. After welding, a steel ram was attached to the end of the cable and pulled through the pipe bank, in turn, easily fracturing back each ring. The broken parts were then flushed out with water.

Avoid skip welding. Skip-welding, a technique of making short, evenly spaced beads of weld metal, is specified many times in construction. Design specifications frequently call for 2-in. (51-mm) welds on 6-in. (152-mm) centers. Figure 3.11 exhibits skip welds in fillet and butt joints. This practice, justified on the basis of saving weld metal and labor, can lead to accelerated corrosion at the inherent crevices in corrosive atmospheres.

To eliminate this problem, the designer can specify continuous light welds, which leave no crevices on the top of the weldment. Actually, the cost will not be a great deal more than skip welding.

Crevices in Vertical Tank Supports

The design of tank supports is very important because of inherent crevices that can foster corrosion. One common design uses concrete bases to support flat-bottom vertical tanks. [See Figure 3.12(a).] Spilled corrosives and water can collect under the tank and severe attack on the tank bottom can result. For example, Degnan reports that the bottoms

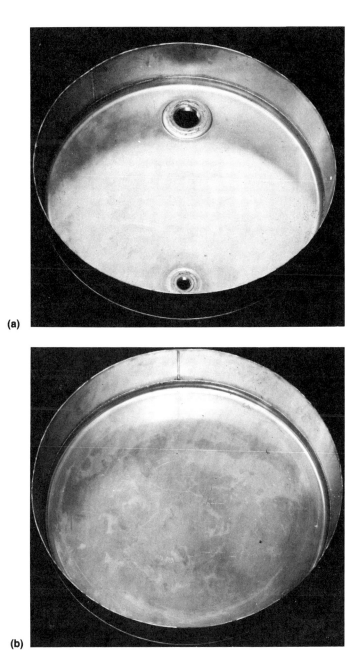

Figure 3.7 — *Inside view of a 55-gallon (208-liter), stainless steel drum, showing full penetration welds without crevices; (a) top and (b) bottom views.*

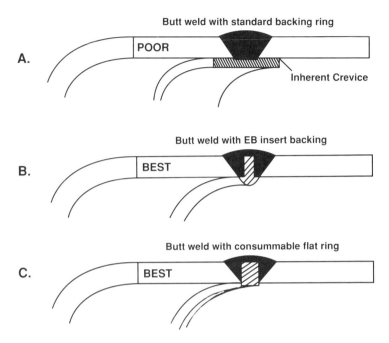

FIGURE 3.8 — *Various ways in which the designer can eliminate crevices by using consumable insert rings.*

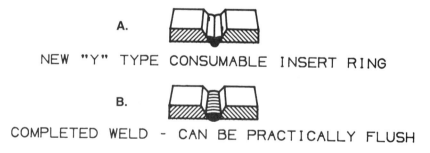

FIGURE 3.9 — *Example of a Y-type consumable insert ring that eliminates crevices.*

of two stainless steel tanks failed after only 60 days of service.[16] Brackish water had seeped under the tanks and chloride stress

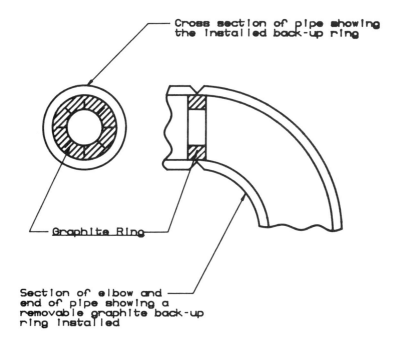

FIGURE 3.10 — *Example of a removable graphite back-up ring.*

corrosion cracking resulted when hot process solutions were stored in the tanks. The tanks had to be completely rebuilt.

In another instance in Niagara Falls, the bottom of a water reservoir for a gas holder made of steel had to be replaced three times over a period of 15 years, while the walls remained in serviceable condition. This corrosion of tank bottoms can be eliminated by the designer if he specifies that I-beams support the tank above the concrete. This eliminates the crevice and allows the tank bottom to be ventilated. The installation of drip skirts also constitutes good design by preventing the drips coming down the tank wall from getting under the tank. [See Figure 3.12(b).]

Sometimes, coal tar or asphalt-saturated felt is placed between the tank bottom and the concrete. This is not entirely satisfactory, as decomposition occasionally causes the felt to become porous and absorb water, which can result in crevice corrosion. Sand is sometimes placed under large tanks primarily to help with leveling. The sand, even when mixed with oil, does not eliminate the crevice corrosion problem.

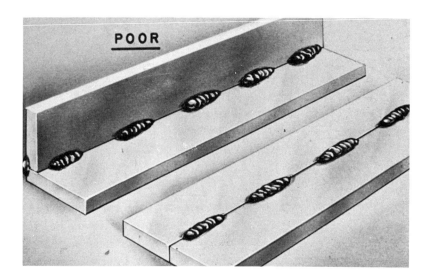

FIGURE 3.11 — *Intermittent or skip welds like those shown above constitute poor design in corrosive service because of the inherent crevice.*

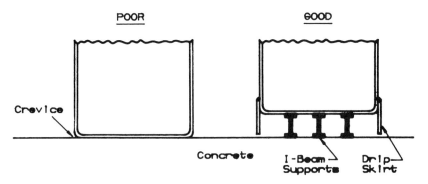

FIGURE 3.12 — *(a) Poor and (b) good tank support designs.*

Horizontal Tank Supports

Horizontal tanks supported on saddles also cause a crevice problem. It is poor design, for example, to support a tank directly on a concrete saddle because of the inherent crevice. Sometimes, a nonporous caulking material is used to fill up the crevice, which is satisfactory if the caulking material does not become porous, as mentioned before.

However, the design is improved if a metal pad of the same material as the tank is placed between the concrete support and the tank and is welded to the tank by continuous filler weld. [See Figure 3.13(a).] The best design, however, is a metal saddle welded to the pad and bolted to the concrete support.[17] [See Figure 3.13(b).]

Tank support design that eliminates crevices is particularly important for stainless steel and aluminum tanks. These materials depend on surface films for corrosion protection. Where crevices exist, these films can be destroyed and cannot rebuild themselves because of the local environmental conditions in the crevice areas.

Crevices in Heat Exchangers

Usually, heat exchangers are fabricated by rolling the tubes into tube sheets to stabilize them and to eliminate, as much as possible, the crevices between the tubes and the top and bottom of the tube sheet. To assure tightness of the joint's inherent crevice on the shell as well as on the bonnet side, the designer should specify the number of grooves to be machined inside the tube sheet hole, the original clearance between the tube sheet hole diameter and the tube diameter, the amount of permanent expansion required in rolling, and the rolling procedure. Such information should be a major part of the basic heat exchanger specification.

The importance of proper rolling cannot be overemphasized. Figure 3.14 shows a portion of an AISI 316 stainless steel heat exchanger tube that had corroded excessively in the crevice area on the shell side of the tube sheet because the tube had not been rolled completely through the tube sheet thickness. Consequently, a substantial crevice was allowed to remain. Hot phosphoric acid on the shell side was the corroding agent. Note in Figure 3.14 that the crevice corrosion progressed almost a third of the way through the tube sheet, while the rest of the tube was practically unaffected. In this particular case, the tube wall was so thin in the corroded areas that all of the tubes had to be replaced, although the remaining lengths of the AISI 316 stainless steel tubes appeared almost new.

Over-rolling can be almost as hazardous as under-rolling. Figure 3.15 shows a sketch of an over-rolled tube. Such over-rolling can weaken the wall and cause heightened corrosion at the over-rolled area.

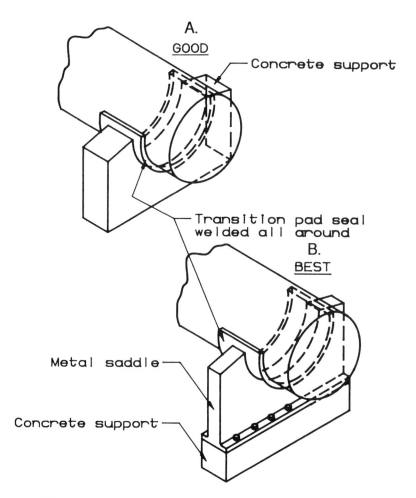

FIGURE 3.13 — *(a) Good and (b) best horizontal tank supports.*

Of course, the most efficient way of eliminating the crevice corrosion problem in heat exchanger tubes is to weld the tube to the tube sheet. This subject will be discussed in detail in Chapter 4.

The baffles in heat exchangers afford crevices in the areas where the tubes pass through. Figure 3.16 shows a piece of AISI 304 stainless steel tube used for a heat exchanger using well water as the coolant. The former location of two baffles can be determined by the clusters of pits evenly spaced about 5-in. (127-mm) apart. The pitting occurred because foreign material (dirt, leaves, debris, etc.) was lodged between

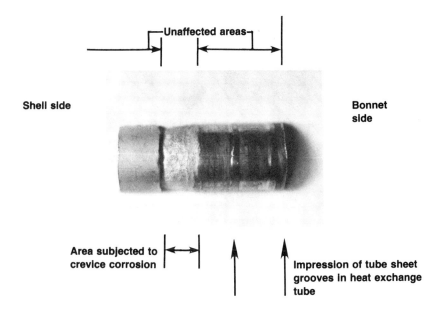

FIGURE 3.14 — *Piece of AISI 316 heat exchanger tube that failed by crevice corrosion; hot phosphic acid was on the shell side.*

the tube and the baffle, thereby setting up crevice corrosion. To avoid this, the designer should leave a generous clearance between the tube wall and the hole in the baffle. Just a small amount of bypassing of the solution through the baffle hole will keep the area clean.

Crevices in Mixer-Settlers

Crevices can cause severe corrosion problems in other types of operating equipment as well. For example, in one large plant, a series of in-line mixer-settlers were installed to handle 30% nitric acid at 160 F (71 C). Both the impellers and their shafts were made of AISI 304L stainless steel. Laboratory tests had shown that this type of stainless steel was very resistant to the corrosive condition. The special design for the impellers required that the impeller be installed last. Because of insufficient space, the impellers could not be welded to the shafts on the top side, so the impellers were bolted to the end of the shaft, as shown in Figure 3.17. After only a few weeks, the shafts and impellers had to be discarded because corrosion had occurred in the crevices existing between the shafts and the impellers, thereby attacking the keys as well as the areas between the nut and the threaded shaft. The first evidence

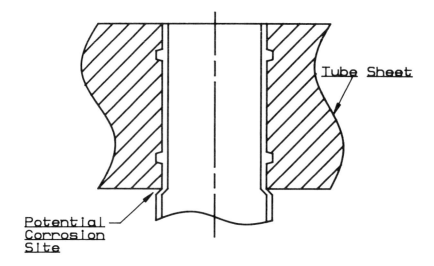

FIGURE 3.15 — *Over-rolled tube into tube sheet can foster local corrosion.*

FIGURE 3.16 — *Pitting on AISI 304 stainless steel tube at baffle locations caused by crevice corrosion.*

of corrosion trouble was when the nuts simply fell off of the shafts. The failure was diagnosed as resulting from the buildup of hexavalent chromium in the crevices between the impeller and the shaft and also between the nut and shaft. Tests[18] of various ions that might accumulate in nitric acid from corrosion of stainless steel have shown that chromium is the principle offender in causing high corrosion rates. Cr^{+3} is oxidized to Cr^{+6} by the acid, and when its concentration reaches about 0.004%, aggressive crevice attack will occur even in annealed, stabilized, or extra-low carbon stainless steel. As the concentration of Cr^{+6} increases, the corrosion rate is increased, which, in turn, produces more Cr^6 and so on. The crevice in this example provided an ideal location for development of such an aggressive condition.

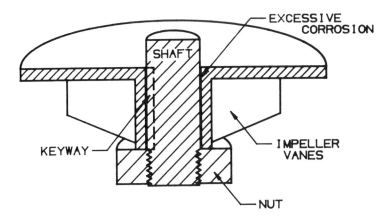

FIGURE 3.17 — *Schematic of an impeller design that failed by corrosion.*

A new design solved this problem by opening the space between the shaft and the impeller hub, as shown in Figure 3.18. Two vent holes were drilled in the impeller hub so that the acid would circulate freely and no buildup of hexavalent chromium would occur. The shaft was welded to the impeller as the last operation. To maintain precise spacing required for dynamic balance of the rapidly rotating impeller, a carbon steel spacer was positioned in the necessary space between the shaft and the impeller before welding. The spacer, of course, could not be physically removed after the welding operation. However, it was rapidly corroded by the acid during the break-in period and was soon completely eliminated. Impellers of this basic design have been in service for years with no further corrosion problems.

Crevices Between Metals and Nonmetals

Contact of a metal to nonmetals, such as plastics, cork, wood, and rubber, can also create crevices. Gaskets perhaps cause the most problems. As long as the gasketed joint is tight and does not absorb water, crevice corrosion will not start.

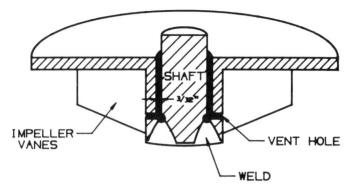

FIGURE 3.18 — *Schematic of an impeller design that experienced a long service life.*

Where gaskets are used on alloys that require oxygen to preserve a protecting thin oxide film, such as stainless steel and titanium, loose gaskets can cause rapid attack. This is because an adequate supply of oxygen is required to rebuild the oxide in breaks in the skin and to preserve passivity, as discussed previously. The area with restricted oxygen becomes the anode and corrodes, while the areas with free access to oxygen become cathodic. Therefore, the designer should carefully scrutinize gaskets that are to be used for such equipment as to strength, stability, lack of "memory," ability to withstand compressive stresses without excessive creep, and resistance to the intended process conditions. Teflon,† Viton,† and specially compounded rubber gaskets have been used with much success. These gaskets are essentially nonwicking, nonhygroscopic materials that will prevent moisture inclusion. Felt, leather, cork, asbestos, or glycol-impregnated gaskets should be avoided. Insulation, wood, cloth, or substances that will absorb water should also not be used in contact with metal.

What the Designer Can Do to Prevent Crevice Corrosion

A careful study of designs by the designer can detect crevices, which often can be avoided entirely or at least minimized. The following are remedies that may be used:

1. Specify that crevices be welded shut.
2. Improve the fit of parts.

---

†Registered trade name.

3. Specify that the crevices be filled with a plastic or an elastomer or other nonporous materials.
4. Change the design to entirely eliminate the crevice.
5. Specify double-butt or double-lap weld joints when practical and possible.
6. When single-butt joints must be used for critical pipelines, consider using consumable or removable inserts.
7. For corrosive environments, specify continuous welds instead of skip welding.
8. Be especially careful when designing tank supports (especially for aluminum or stainless steel tanks) to assure as few crevices as possible.
9. Seal weld tubes to tube sheets when practical. If seal welding cannot be specified, make sure that the tube installation procedure will result in a tight fit of tubes to tube sheet.
10. In critical rotating equipment, where crevices cannot be eliminated, open the crevices enough so that they will circulate the solution, thus avoiding crevice corrosion problems.
11. Do not specify that any material which absorbs water be placed next to metals or alloys.

◻ 3.2 Galvanic Corrosion

The name for a separate form of corrosion, *galvanic corrosion*, is rather confusing because all corrosion cells are galvanic in nature, as discussed previously. *Galvanic corrosion* actually means the corrosion generated between two different metals. The more appropriate name would be *two metal corrosion*, as has been suggested by several authors. When designing equipment, it is not always feasible, or, for that matter, desirable to specify that all the parts be made from the same metal or alloy. As mentioned before, when different metals are used together in a media that will carry an electrical current, the more active of the metals (the anode) may be corroded. Whether corrosion will occur or not depends on the relative activity of the metals involved.

An interesting experiment was conducted many years ago in 1761, which first involved the study of galvanic corrosion. A report[19] by the British Navy Board concerned the first "coppering experiment" where the hull of the British frigate HMS Alarm was sheathed in 12-oz copper sheeting in an attempt to stop the deterioration of the hull by ship worms (*toredo*). The copper sheets were secured to the hull by iron nails. When the frigate returned from the trial run to the Azores, some of the sheets were gone and others were barely hanging on. The reason for the failure was that the iron nails in contact with the copper had corroded.

The report, dated August 31, 1763, clearly disclosed that the investigators understood that certain combinations of metals in contact could cause corrosion of one of the metals and that insulating one metal from another could stop such aggressive action.

If the information in the report had been seriously applied by others at that time, the knowledge of galvanic corrosion could have been greatly advanced, averting many decades of economic loss.

Galvanic Series of Metals

A galvanic series may be compiled based on the potentials (voltages) of various groupings of metals in a given electrolyte. The potential of a metal in a solution is related to the energy that is released when the metal corrodes. The metals are arranged with the most active (less noble, more anodic) at the top of the list and with the least active or passive (more noble, more cathodic) below. The higher position of a metal in the series indicates that it will show accelerated corrosion when coupled with a metal below it. A galvanic series, however, does not show the *magnitude* of the corrosion between such couples. This is because while the galvanic series is based on potential differences, actual corrosion is based on electrical current *flow*. Faradays' Law states that the current is directly proportional to the quantity of metal removed by the corrosive attack. For instance, one ampere of electrical current flowing in one year can remove about 20 lb (9 kilograms) of steel.

Table 3.1[20] shows a galvanic series in seawater flowing at 8 to 13 feet/second (ft/s) (146.32 to 237.77 m/min) at a temperature range of 50 to 80 F (10 to 27 C). A calomel half-cell is used as a reference. Most observations have been made on the galvanic behavior of metals in seawater, however, other media have been tested. For instance, Table 3.2[21] displays a galvanic series based on the atmosphere as the media, and Table 3.3[22] uses plant conditions as the test media.

The galvanic series can be very useful to the designer, even though the positions of metals can change somewhat with varying environmental conditions and temperature. Note the change in the galvanic positions of various metals in Tables 3.1 through 3.3. The designer, however, may use these tables for initially determining safe couples for other environments. These candidates should then be tested in the actual process media according to the techniques covered in Chapter 2.

TABLE 3.1
Corrosion ——— Potentials in Flowing Seawater (8 to 13 Ft/S) Temperature Range of 50 to 80 F[1]
VOLTS: SATURATED CALOMEL HALF-CELL REFERENCE ELECTRODE

Alloy	Potential Range (V)
Magnesium	~-1.6
Zinc	~-1.0
Beryllium	~-1.0
Aluminum Alloys	~-0.8 to -1.0
Cadmium	~-0.7
Mild Steel, Cast Iron	~-0.6 to -0.75
Low Alloy Steel	~-0.6
Austenitic Nickel Cast Iron	~-0.45 to -0.6
Aluminum Bronze	~-0.3 to -0.4
Naval Brass, Yellow Brass, Red Brass	~-0.3 to -0.4
Tin	~-0.3
Copper	~-0.3
Pb-Sn Solder (50/50)	~-0.3
Admiralty Brass, Aluminum Brass	~-0.3
Manganese Bronze	~-0.3
Silicon Bronze	~-0.3
Tin Bronzes	~-0.3
Stainless Steel - Types 410, 416	-0.25 (active shift to ~-0.5)
Nickel Silver	~-0.25
90-10 Copper-Nickel	~-0.25
80-20 Copper-Nickel	~-0.25
Stainless Steel - Type 430	-0.2 (active shift to ~-0.5)
Lead	~-0.2
70-30 Copper-Nickel	~-0.2
Nickel-Aluminum Bronze	~-0.2
Nickel-Chromium alloy 600	-0.15 (active shift to ~-0.5)
Silver Bronze Alloys	~-0.15
Nickel 200	~-0.15
Silver	~-0.1
Stainless Steel - Types 302, 304, 321, 347	-0.1 (active shift to ~-0.5)
Nickel-Copper alloys 400, K-500	~-0.1
Stainless Steel - Types 316, 317	-0.05 (active shift to ~-0.5)
Alloy "20" Stainless Steels cast and wrought	~0
Nickel-Iron-Chromium alloy 825	~0
Ni-Cr-Mo-Cu-Si alloy B	~0
Titanium	~+0.05
Ni-Cr-Mo alloy C	~+0.05
Platinum	~+0.25
Graphite-carbon	~+0.25

Passive ←————————————————→ Active
Most Noble ←————————————————→ Least Noble
Cathodic ←————————————————→ Anodic

(1) Alloys are listed in the order of the potential they exibit in flowing seawater. Certain alloys indicated by the solid bar symbol ▬ in low-velocity or poorly aerated water, and at shielded areas, may become active and exhibit a potential near -0.5 volts. Shift indicated by arrow ⌇►.
Reference 20

TABLE 3.2
Galvanic Series in the Atmosphere[1]

Group	Composition
I	Magnesium
II	Aluminum
	Zinc
	Cadmium
III	Iron and Carbon Steels
	Lead
	Tin
IV	AISI 430 stainless steel
	AISI 302 stainless steel
	AISI 304 stainless steel
	AISI 316 stainless steel
V	Copper—Nickel and Copper-Zinc Alloy[2]
	Copper[2]
	Silver[2]
	Gold

[1] Based on atmospheric tests in marine, industrial, and severe tropical atmospheres.[21]
[2] Out of order compared to Table 3.1.

Active and Passive Conditions

Certain alloys, indicated by the solid bar symbol in Table 3.1, may shift from a passive to an active condition, thereby changing their relative position in the table. (Such shifts in position are noted in Table 3.1 by arrows.) As discussed previously in Chapter 1 and earlier in this chapter, some metals and alloys need oxygen for corrosion protection. Stainless steels, for instance, depend on oxide film for protection and are aided by the presence of oxygen. When the film is intact, the surface is called *passive*. When the oxygen content is high, the oxide film is maintained and corrosion resistance is high. When the oxygen concentration is low, breaks in the oxide film are regenerated with difficulty and higher corrosion rates result. When the protective film has been removed, the surface of the metal is called *active*. Refer to Chapter 1.4.

TABLE 3.3
Galvanic Series Based on Numerous Tests Under Plant Conditions[1]

Corroded End (Anodic or Least Noble)	For Comparison: Grouping in Table 3.2
Magnesium Magnesium Alloy	I
Zinc	II
Aluminum	II
Cadmium	II
Aluminum 2017	—
Steel or Iron Cast Iron	III —
Chromium-Iron (active)	IV
Ni-Resistant	—
AISI 304 SS (active) AISI 316 SS (active)	IV
Lead-Tin Solder Lead Tin	— [2]III [2]III
Nickel (active) Inconel (active) Hastelloy C (active)	IV — —
Brasses Copper Bronzes Copper-Nickel Alloys Monel	V V — V —
Silver Solder Nickel (passive) Inconel (passive)	—
Chromium-Iron (passive) AISI 304 SS (passive) AISI 316 SS (passive) Hastelloy C (passive)	
Silver	
Graphite	—
Gold Platinum	V —
Protected End (Cathodic or Most Noble)	

[1]From Reference 22.
[2]Out of order compared to Table 3.2.

Relative Areas of Anodes and Cathodes

The use of metals or alloys listed far apart in Table 3.1 should, in general, not be used in an atmosphere where a current between the two members of the couple can flow. However, the designer should always consider the relative areas of the cathode and the anode before selecting dissimilar metals. For instance, if a copper tank contains steel rivets, the rivets will rapidly corrode when the tank contains brine, as Table 3.1 shows. However, if the tank was made of steel and the rivets of copper, the tank would last for many years. The reason for this is that the area ratio of the anode and the cathode determine current density. In the first case, the cathode area was large, and therefore a high current density was imposed on the small total area of the steel rivets (the anode). In the second case, the copper rivets (the cathodes) were small in total area, thus the current density was small. The low current was then imposed over a large anodic area, which resulted in very low corrosion rates. In each case, the electrical potential difference (voltage) between the copper and the steel is the same. This potential difference is what the galvanic tables are based on. However, the difference in current densities at the anode result in different corrosion rates.

A rule-of-thumb the designer may use in designing for dissimilar metals is called the *Area Ratio Rule*, which states that the rate of corrosion will vary directly with the increase or decrease in the area ratio of the more noble metal to the less noble metal when connected together in an electrolyte. For example, if the anode is half the area of the cathode, the corrosion rate will be doubled compared with the corrosion rate when both areas are the same. Conversely, if the anode area is double that of the cathode, the corrosion rate will be half.

The area rule only applies when the corrosion reaction is controlled by the cathode. To explain, during normal corrosion, the electric current flows from the anode into the solution and hence to the cathode with the cathode controlling the reaction. When there is anodic control, the area rule does not apply. (See Chapter 6.4.)

Galvanic Series

When the designer has to select dissimilar metals for weathering conditions, Table 3.2 should be consulted. In developing this galvanic series, many tests were conducted in marine, industrial, and severe tropical atmospheres to determine practical corrosion information. Any two materials within any one of the five groups of Table 3.2 may be considered safe from galvanic corrosion when coupled together in an electrolyte. If the two materials are from different groups, galvanic corrosion may occur.

As will be noted in Table 3.2, copper-nickel, copper-zinc alloys, and copper are out of order compared to that of Table 3.1, which summarizes results from tests conducted in running seawater. These three metals are much more noble in the atmosphere than in running seawater. Galvanic action in the atmosphere does not extend very far from the area of contact of the dissimilar metals because of the high resistance of the thin films of the electroylyte that can form. (This fact probably accounts for the differences between Tables 3.1 and 3.2.) The area where corrosive action occurs on the anode between the couples is very narrow, only about one-half inch or less in width. If this area is occupied by an inert washer, such as plastic or rubber that can seal off the area from the electrolyte, galvanic action can be stifled.

Another helpful series is shown in Table 3.3. This table is based on actual corrosion testing experience with numerous corrosives in the laboratory under plant operating conditions and on practical results with metals and alloys in long service. There are 17 different groups in this table. Any coupling of metals within a single group is considered safe. The coupling of metals from two different groups may cause corrosion of the member that is more anodic or higher in the list than the other members.

Insulation of the Anode from the Cathode

Since there are differences in the various galvanic series, the safe procedure for the designer is to specify that the anode and cathode be insulated from each other, as shown in Figure 3.19(a). In each of the three examples, the electrical path is blocked so no galvanic accelerated electrical current flow can exist and therefore no galvanic corrosion.

The following are examples of good galvanic design:

> *Example 1* — The potential galvanic corrosion of a corrugated aluminum roof was avoided by a unique design[17] depicted in Figure 3.20. Special welding studs were welded to the steel purling. The studs consisted of steel shells with aluminum shoulder inserts. The aluminum roof was installed over the aluminum shoulders, the rubber grommets were positioned, and the roof was riveted into place.

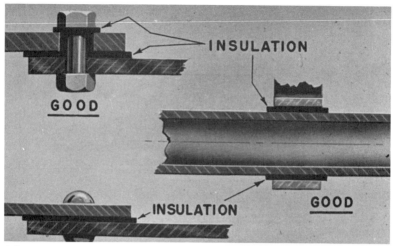

(a)

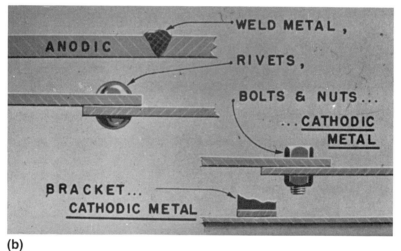

(b)

FIGURE 3.19 — *(a) For dissimilar metals, insulate one from the other; (b) for dissimilar metals that cannot be insulated, make the more noble metal (the cathode) the smaller area.*

> *Example 2* — When less noble bolts are used in the weather for more noble pipe flanges (larger area), such as high-strength carbon steel bolts for stainless steel flanges, the bolts can

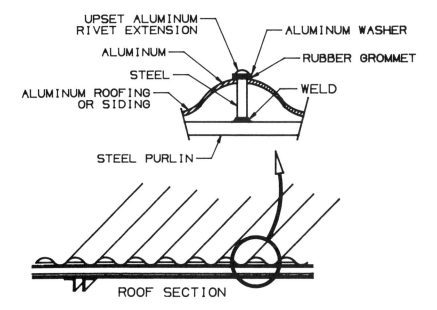

FIGURE 3.20 — *Example of a good roof design used to avoid galvanic corrosion.*

fail by galvanic action. (Stainless steel bolts may have too low a yield strength to ensure tight closure in some applications.) Dissimilar metals must therefore be insulated from each other. A sleeve should be installed over the bolts with washers at both ends, as shown in Figure 3.21, to assure no electrical current flow between the parts of the couple.

When Insulation is Impractical

There are many environments in which insulation is not feasible, such as high temperature or aggressive solutions where insulation will deteriorate. In these cases, the more noble metal should be used for joining or fastening the less resistant metals.

There have been many applications of dissimilar couples where excellent service life has been realized, even though a study of galvanic series chart would indicate that such couples might fail. The reason for the success was that the designer had followed the Area Ratio Rule and had made the anodic side of the couple much larger in area than the

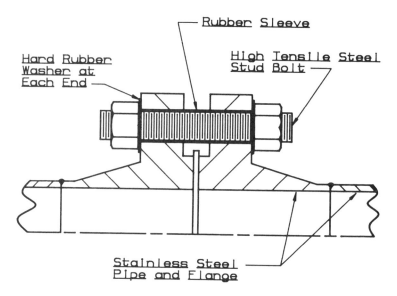

FIGURE 3.21 — *When the anode [the high tensile steel stud bolt] is smaller than the cathode [the stainless steel flange and pipe], insulate with nonmetallic sleeve [rubber sleeve] and washers [hard rubber washer at each end], as shown above.*

cathodic side of the couple, as exhibited in Figure 3.19(b). In each form of joint depicted, whether welded, bolted, riveted, or clamped, the anode is many times larger than the cathode. Practical examples seen in industry are:

1. Iron or cast iron valve bodies (anode, larger area) trimmed with austenitic stainless steel or Stellite[†] (cathode, smaller area).
2. Brass fittings used with carbon steel pipe.
3. Stainless steel bolting on aluminum alloy equipment.
4. Monel[†] bolts used to hold together parts of steel condensers exposed to brackish cooling water.[20]
5. Monel bolts or rivets used to secure removable iron impellers in iron pumps handling strong electrolytes.
6. Water heaters with copper tubes installed into steel or cast iron tube sheets are in fairly common practice, in spite of what the galvanic series might indicate. This practice avoids the higher cost of bronze tube sheets.

[†]Registered trade name.

Protective Coatings for Galvanic Couples

Protective coatings or paints can be specified to insulate couples from each other, but this can sometimes be a rather dangerous procedure. For example, the less noble materials should not be painted without also coating the more noble materials.

An example[23] of what happens when this is not done involved milk coolers. The tank for these particular coolers was made of AISI 316 stainless steel, while the reservoir for the cooling brine was steel coated with epoxy paint. (See Figure 3.22.) After the manufacturer had sold a large number of these units, an epidemic of leaks was reported. At first, sabotage was suspected, since small holes appeared to have been drilled through the painted brine reservoir. However, this "drilling" had actually been caused by corrosion. The large AISI 316 milk tank had acted as a cathode, and the inherent small pinholes in the paint film had exposed very small areas of steel that acted as anodes. Because of the high current density, the pinholes rapidly penetrated the light gage steel. (This principle is used in electrolytic drilling.) Painting the tank (anode) without also painting the milk tank was a mistake because the anodic area was drastically reduced, while the cathodic area remained large. The obvious solution was to make the brine reservoir of a material closer to stainless steel in the galvanic (seawater) series. Since this would have been too expensive, the vendor compromised by painting the bottom of the milk tank, which reduced the effective cathodic area and thus prolonged the life of the coolers. The manufacturer was happy because his complaints ceased, at least for awhile, after the coatings had been applied.

Unusual Cases of Galvanic Corrosion

The following examples represent unusual cases of galvanic corrosion:

Plant layout. During the early phases of plant design, when the equipment layout is being studied, the designer should realize that galvanic couples can also be set up between interconnecting vessels of dissimilar metals if there is a common electrolyte and the vessels are supported on the same steel support. It has been reported[17] that an aluminum mixing tank was corroded severely when connected to a copper process vessel when both units were supported on a common steel frame. The aluminum was the anode. Galvanic current flowed from the aluminum mixing tank through the electrolyte to the copper vessel (the cathode) and then back through the steel support to the aluminum.

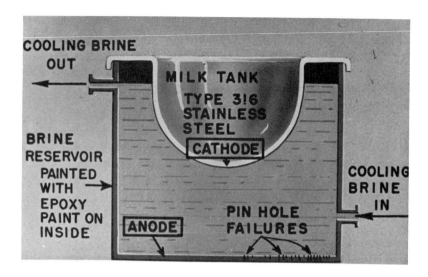

FIGURE 3.22 — *Poorly designed milk cooler that failed by corrosion.*

Graphite and carbon dangers. The designer should be cautioned about the use of graphite or carbon. Table 3.1 shows graphite-carbon to be more noble than the metals listed, even below platinum. Therefore, carbon will form a galvanic couple with most materials in an electrolyte.

The following examples will clarify this hazard:

Example 1 — Magnesium sheet was deep-drawn into drums used to contain concentrated hydrofluoric acid. After the acid was used by the customer, the drums were washed out and sent back to the acid supplier. When they arrived, the drums were badly corroded, containing many perforations. The magnesium withstood the aggressive hydrofloric acid very efficiently; however, graphite had been used by the drum manufacturer as a lubricant in the deep-drawing operation. Some graphite was embedded in the surface of the magnesium. The galvanic couple between the magnesium (anode) and the graphite (cathode) in the presence of moisture caused the perforations.

Example 2 — Aerospace equipment has been plagued by galvanic corrosion because of the presence of carbon and graphite. For this reason, solid film lubricants are not allowed for use unless they are entirely free of powdered carbon or graphite.[24] Cadmium-plated fasteners are also prohibited for use in contact with graphite composites.

Example 3 — In another instance, carbon Raschig rings were used in a AISI 347 stainless steel absorption tower. After about a year of operation, it was found that the inside of the tower was deeply pitted. The pits appeared wherever the Raschig rings contacted the stainless wall. This contact in an electrolyte resulted in galvanic corrosion. The tower was rebuilt and ceramic rings used instead of carbon Raschig rings. Afterwards, there was no galvanic corrosion problem.

Example 4 — A copper pipe was connected to an aluminum vessel. Copper ions were evolved from the copper pipe by oxidation ($Cu^o \rightarrow Cu^{++}$) and traveled inside the vessel. Copper was then precipitated on the wall of the aluminum tank where galvanic corrosion caused holes to be formed damaging the tank.

Example 5 — In yet another instance, a steel pipeline, buried in the ground, was pitted from the outside in. An investigation disclosed that the ground where the pipeline had failed showed the remnants of a coal pile that used to be located there. In the moist ground, galvanic cells were set up between the residual carbon and the carbon steel. A section of the pipeline had to be replaced.

Effect of Size on Polarization

Another factor controlling corrosion not reflected in potential differences between couples is polarization, which was discussed previously in Chapter 1.4. Where there is a large cathode and a small anode, for instance, there is very little polarization to reduce the galvanic effect. Where there is a small cathode and a large anode, however, this condition is conducive to polarization, which curtails corrosion. Therefore, the final decision on the selection of couples for design should be based on actual tests of galvanic corrosion specimens simulating the difference in the areas of the anode and cathode conducted in the expected corrosive environment.

Danger of Heavy Metals

Heavy metals, such as mercury, can cause unexpected galvanic corrosion. To cite an example, a plant in New Jersey produced formaldehyde using aluminum processing equipment, including towers containing bubble cap trays. The aluminum equipment withstood the process conditions for years with no corrosion problems when suddenly, the process yield dropped dramatically and it was discovered that some of the aluminum equipment had been badly corroded. Figure 3.23 shows a badly pitted bubble cap from that plant. In some areas, massive pits penetrated through the aluminum. An investigation disclosed that several manometers had lost mercury because of a process pressure upset. As a result, mercury was broadcast inside much of the aluminum equipment. The lost mercury formed many cathodes while the aluminum acted as the anode and aggressive galvanic cells were formed, which caused the damage.

FIGURE 3.23 — *Aluminum bubble cap from aluminum towers producing formaldehyde. (Mercury from a manometer was responsible for this massive corrosion attack.)*

What the Designer Can Do to Prevent Galvanic Corrosion

There are many actions a designer can take to prevent galvanic corrosion. Various suggestions are outlined in the following list:

1. Review all designs to assure that no hazardous galvanic couples have been inadvertently specified.//
2. Use galvanic series tables to select suitable couples. Confirm these by laboratory tests.

3. When possible, electrically separate the dissimilar metals with an electrical insulator.
4. When insulation is not practical, follow the area-ratio rule and specify that the area of the anode be much larger than the cathode.
5. When specifying that paint be used to insulate one member of a couple from the other, never just specify painting the anode only. Either paint both the anode and the cathode or just the cathode.
6. Study plant layouts and separate electrically any dissimilar metal process equipment that hold solutions and are connected.
7. When absorption tower packing is to be specified, make sure the material of construction of the packing does not form a galvanic couple with the tower walls.
8. Insulate buried steel pipe to obviate any cathodic affects from the soil.
9. Do not specify manometers and other devices containing mercury for monitoring process equipment made of any metal high in the galvanic series. If this is not practical, specify catch pots to trap the mercury.

☐ 3.3 End Grain Attack

Definition of End Grain Attack

In severely corrosive environments, such as boiling nitric acid, austenitic stainless steels are susceptible to a form of selective corrosion termed *end grain attack*. This form of attack is sometimes observed. *End grain* is defined as the surface of the metal which is cut perpendicular to the direction of rolling. During the rolling of stainless steel at the mill, nonmetallic inclusions, inherently present in the steel, are elongated in the direction of rolling.

As the ingot is rolled, the ingot is successively broken down (rolled) to smaller and smaller thicknesses. During such action, nonmetallic inclusions are rolled out in thinner and thinner sections. When the sheet or plate has been rolled to the proper gage, the nonmetallic inclusions have been rolled into long "stringers," all in the direction of rolling, as shown in Figure 3.24.

When the end grain is exposed to a strong corrodent, end grain attack starts at areas where stringers are exposed on sheet or plate ends and can aggressively proceed down into the steel. Stainless steel sides and surfaces, on the other hand, do not corrode.

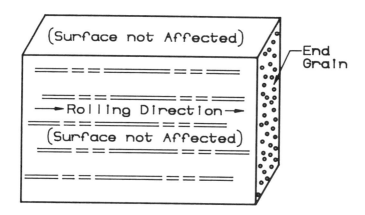

FIGURE 3.24 — *Schematic of a stainless steel plate showing long lines of inclusions and stringers.*

Another way of explaining this is to consider holding a bundle of soda straws in your hand. The open ends of the soda straws are analogous to the open ends of the rolled out impurities or stringers, while the sides of the straws, on the outside of the bundle, are analogous to the flat and side surfaces of the sheet or plate. If the end grain of the stainless steel is subjected to a strong corrodent, corrosion cells can easily be set up. Such corrosion can continue down the length of the stringers, forming deep pits. This is like going down the ends of the soda straws. On the contrary, the flat surfaces and sides (the length of the straws) when confronted with a corrodent, do not form deep pits because of the very small diameter of the stringers and inclusions.

Stainless steel piping and bar stock are also susceptible to end grain attack.

Carbon steels may also be susceptible to end grain attack. The most susceptible are the resulfurized, free-machining grades such as AISI C 1120 and C 1132, which contain large amounts of sulfides which become elongated in the rolling direction. These and other second phases can trigger end grain attack.

Examples of End Grain Attack

The two following paragraphs give examples of end grain attack.

Example 1 — In most instances of fabricated equipment, the end grain is covered up. For instance, when a drum is constructed by rolling a sheet or a plate into a cylinder, a longitudinal joint is formed and welded. Then, top and bottom heads are welded on. Consequently, the end grains are covered up by the welds obviating any end grain attack. However, there are some cases where end grain is left exposed, as in bubble caps, riser pipes, sieve plates in absorption columns, and filler or open-ended feeder pipe. Bored holes and slots in vessels and piping are also susceptible to this form of attack. An example of end grain attack was shown in an situation where AISI 304L stainless steel rotating drums containing many round holes were used in handling a strong corrodent. It was observed that the round holes were gradually assuming an eliptical shape during the months of production. The longer the time was, the more elongated the holes became. What had happened was that the corrodent had attacked the exposed end grain, leaving the rest of the interior surface of the hole relative untouched. (See Figure 3.25.)

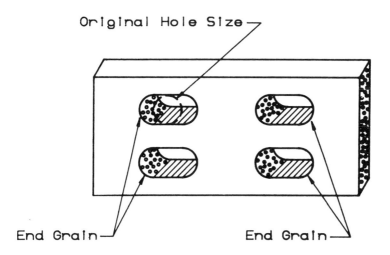

FIGURE 3.25 — *Elongation of holes in stainless steel plate at exposed end grain.*

Example 2 — In another example where bubble cap distillation columns were used in hot strong waste nitric acid service, the bubble caps and vapor risers were literally eaten up through exposed end grains, while the rest of the columns were practically untouched. (See Figure 3.26.) As shown, many of the bubble caps were completely gone and the risers were severely attacked.

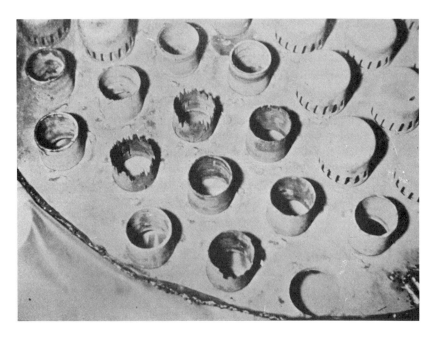

FIGURE 3.26 — *Riser pipes and bubble cap corroded by end grain attack.*

End Grain Attack on Stabilized vs Unstabilized Stainless Steels

Stabilized or extra-low carbon stainless steel is almost as susceptible to end grain attack as sensitized nonstabilized stainless steel, as Figure 3.27 displays. (Chapter 5 will explain the difference between stabilized and unstabilized stainless steel.) Two welded specimens of AISI 304L and 304 had been exposed to boiling 62% nitric acid for 48 hours. Intergranular attack had occurred in the welded AISI 304 specimens (the lower one); note the perforation along each side of the weld. (Intergranular corrosion will be discussed later in this chapter.) However, the specimens were attacked almost equally at the end

FIGURE 3.27 — *Example of end grain attack; this view is from the end grain of welded AISI 304L stainless steel (top) and AISI 304 stainless steel (bottom). (The AISI 304 specimen has also been attacked intergranularly.) These specimens were exposed to boiling 62% boiling nitric acid for 48 hours. (Note that both AISI 304 and 304L stainless steels are susceptible to end grain attack, but not either weld.)*

grains. The other surfaces of the specimens were not attacked in this manner. Note that the welds were also unaffected.

What the Designer Can Do to Prevent End Grain Attack

As the designer reviews his equipment designs, he should be alert to possible exposed end grains in equipment intended for service in an aggressive corrosive environment. He can specify on his drawings exactly where he desires end grain protection. He should use AWS welding symbols for designating the type of weld, as will be discussed in Chapter 5.

The solution to the end grain problem is to "butter" the end grain with weld metal. As noted in Figure 3.27, weld metal is not susceptible to this form of attack.

As shown in Figure 3.28, welding can be applied to the edges of bubble caps or the exposed edges of the riser pipes. Slots in pipes can be coated with weld metal, as also shown in Figure 3.28. Where there are holes in process equipment exposed to aggressive solutions, the interior of the holes should be coated with weld material. There does not have to be a weld deposit with filler rod. Fusing the surface of the end grain with a tungsten arc alone is sufficient to eliminate end grain attack.

Where possible, modify design details so that end grains are not exposed.

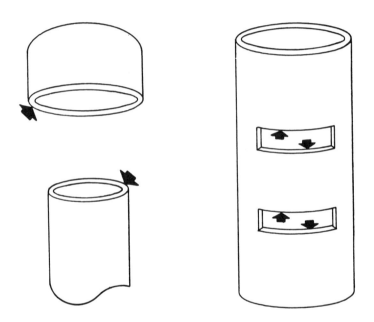

FIGURE 3.28 — *Example of how end grain susceptibility can be eliminated by "buttering" with weld metal at the arrow locations.*

☐ 3.4 Erosion-Corrosion

This form of corrosion involves the simultaneous effects of both erosion and corrosion at high solution velocities. Solids in suspension are often associated with erosion-corrosion attack, but are not necessary to cause this form of attack; however, solids can aggravate the attack by helping to destroy the protective film. Specific types of erosion-corrosion include wire drawing (grooving), impingement, cavitation, and fretting corrosion. Both high and low velocities can alter the corrosive attack of process streams. As process streams increase in velocity, they become progressively more aggressive. Also, when the velocity is practically zero, the corrosion rate can also be dramatically increased. The designer must always take velocity into consideration when designing piping and equipment.

Alloys Aided by Oxygen

Most metals owe their corrosion resistance to a protective film or skin on the metal surfaces. Any mechanical disturbance of this film can lead to increased corrosion. Increasing the solution velocity makes it more difficult to maintain the protective film. Certain metals are not as susceptible to higher velocities when oxygen is present. These alloys need oxygen to maintain their protective films, as discussed previously. For instance, austenitic stainless steels (300 Series), Hastelloy[†] C-276, and titanium depend on a passive oxide film for corrosion protection and hence, are aided by increased oxygen. These alloys make the best pump materials.

Alloys Not Aided by Oxygen

Because they are not aided by oxygen, copper and copper-base alloys, which thrive on reducing conditions while austenitic stainless steels do not, are very sensitive to increasing fluid velocities. There is a critical velocity where the protective film is not replaced as rapidly as it is formed. At that velocity, corrosion rates can increase greatly. There is a significant variation in the maximum permissible (critical) velocity of copper-base alloys in salt water, as noted below:

Copper	2 to 3 ft/s (37 to 55 m/min)
Admiralty Metal	5 to 6 ft/s (91 to 110 m/min)
Arsenical Aluminum Brass	7 to 8 ft/s (128 to 146 m/min)
90 to 10 Cupro-Nickel	8 to 10 ft/s (146 to 183 m/min)

Reasons for the above variations are the differences in composition and tenacity of the various protective films involved. Cupro-nickel alloys,

---

[†]Registered trade name.

for instance, can be used at fluid velocities 4 to 5 times faster than copper because the nickel in the alloy imparts more stability to the protective film.

Effects of Velocity

The effects of velocity are depicted in the following six examples:

> Example 1 — Velocity has a significant effect on the corrosion resistance of carbon steel in seawater at ambient temperature, as exhibited in Figure 3.29. There is a fourfold increase from about 0 to 15 ft/s (0 to 274 m/min).

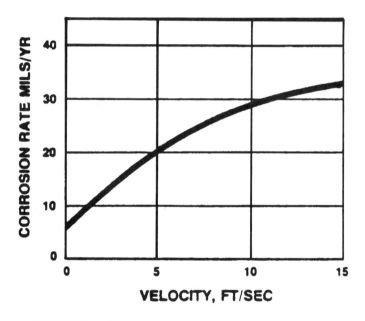

FIGURE 3.29 — *Effect of seawater velocity on corrosion of steel at ambient temperature after a 38-day exposure. (H. R. Copson, Corrosion, Vol. 16, No. 2, 1960.)*

> Example 2 — When acids are involved, the effect of velocity is often very pronounced. As an example, a dramatic failure is shown in Figure 3.30. A centrifigal pump casing failed after

72 hours of service in 7% oleum (fuming sulfuric acid). The pump housing was badly attacked, while the impeller practically disappeared. Steel had been selected as the material of construction for this pump by the designer on the basis of a demonstrated six-year life of steel pipe handling acid of the same strength. The cause of the pump failure was simply an increase in acid velocity. The acid in the pipeline flowed at a rate of only about 3 ft/s (55 m/min), while the centrifugal pump was operating at an acid velocity of well over 80 ft/s (1,463 m/min).

Example 3 — The effects of velocity on corrosion are shown

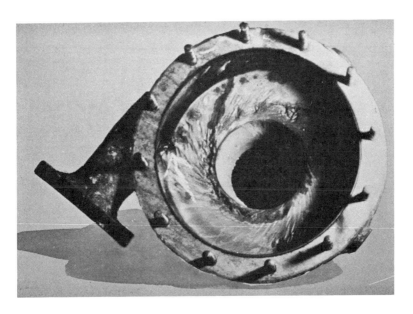

FIGURE 3.30 — *Steel centrifugal pump housing that failed after 72 hours in oleum service (the impeller practically disappeared).*

graphically in Figure 3.31, which exhibits the results of a dynamic corrosion test conducted on five piping materials in 99.6% sulfuric acid at 300 F (149 C). These tests were conducted to select the material of construction and the maximum flow that would be allowed for a chemical plant operation in Cleveland, Ohio. It had been decided that because of a limiting wall thickness, only 3 mils (0.003 in.) per month loss of metal could be tolerated. It is interesting to observe that the corrosion rates of these five metals at zero velocity is below the maximum corrosion rate set. However, as the velocity of the acid was increased, gray cast iron would fail at any velocity; AA[4] aluminum 1100 could be used to about 2 ft/s (37 m/min), Alloy 20 to a little less than 3 ft/s (55 m/min), AISI 316 to a little over 4 ft/s (73 m/min), and AISI 304 stainless steel to about 5 ft/s (91 m/min). Therefore, AISI 304 stainless steel was selected with the acid velocity limited to 5 ft/s (91 m/min). The corrosion test

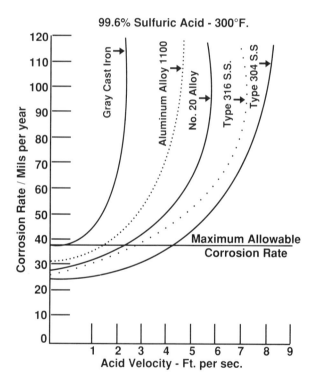

FIGURE 3.31 — *Test results of five materials in 99.6% sulfuric acid at varying solution velocities at 300 F (149 C).*

for the AISI 304 stainless steel was accurate, since this equipment had a long service life.

Example 4 — Another dynamic test was conducted to prove to a customer that inhibited 5% sulfamic acid, when used for cleaning AISI 1020 steel, would remove less metal than inhibited 5% hydrochloric acid. The tests were conducted at various acid velocities. As can be seen in Figure 3.32, the sulfamic acid was far less aggressive at the higher velocities, normally used for cleaning plant piping, than hydrochloric acid. Both acids cleaned AISI 1020 steel efficiently. The rates for hydrochloric acid increase almost linearly with increasing velocity, while sulfamic acid rates level off after 2 ft/s (37 m/min). At 2 ft/s (37 m/min), hydrochloric acid is 4 times more corrosive than sulfamic acid, while at 6 ft/s (110 m/min), the hydrochloric acid has over 9 times the corrosive effect.

Example 5 — In another example, pipelines and facilities were being designed for a chemical plant in New Jersey for handling a waste stream of contaminated 79% sulfuric acid plus organics and water at 100 F (38 C). Figure 3.33 shows that under static conditions, corrosion rates were the same

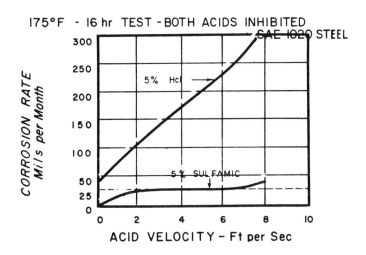

FIGURE 3.32 — Comparison of hydrochloric acid with sulfamic acid in dynamic corrosion tests.

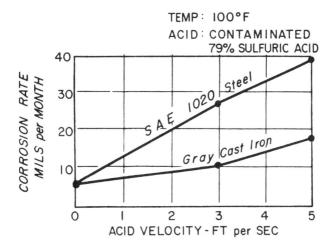

FIGURE 3.33 — *Dynamic corrosion test results of steel and gray cast iron in waste acid.*

for mild steel and cast iron. At a velocity of 1 ft/s (18 m/min), the corrosion rates were fairly close; however, when the velocity was 3 ft/s (55 m/min), the carbon steel corroded over 2.5 times that of the cast iron. When the velocity was raised to 5 ft/s (91 m/min), corrosion rates of both materials were increased signficantly, with the steel being the highest. Actually, carbon steel piping was selected for the job and was sized to provide a maximum flow rate of 1 ft/s (18 m/min). This selection was based on the much lower cost of carbon steel pipe and the ease with which the carbon steel pipe could be installed in the field by welding. By knowing the limitations of the carbon steel under velocity conditions, the designer was able to avoid corrosion problems and thus save both material and installation costs and still have a safe pipeline.

Example 6 — For many piping systems, the designer must balance the extra cost of larger pipe and hence lower velocity for a given through-put, with the increased corrosion at a higher velocity. A plant rule of thumb for replacing carbon steel pipelines that have corroded in a short time in 60° Baume (78%) sulfuric acid, is simply to increase the

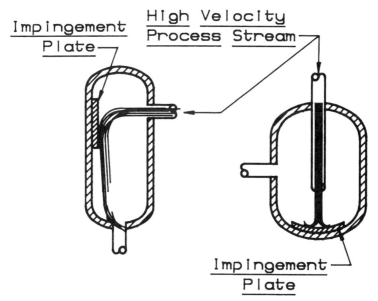

FIGURE 3.34 — *Use of impingement plates.*

pipe diameter by 1/2 to 1 in. (13 to 25 mm). For instance, if a 2-in. (51-mm)-diameter line failed, then the replacement line would be either 2 1/2 or 3 in. (64 or 76 mm) in diameter. This change in pipe size would significantly lower the corrosion rate by lowering the velocity and thus greatly lengthen the life of the pipeline.

Impingement Attack

Another problem that the designer sometimes experiences, is high velocity streams discharging against the wall of a vessel or other process equipment. At the point of impingement, rapid corrosion may occur. The designer has the four following choices to solve this problem:

1. Add reinforcing pads to the area of impingement, as shown in Figure 3.34.
2. Add sacrificial baffles.
3. Redirect inlet streams away from the equipment walls; or
4. Lower the stream velocity.

Localized High-Velocity Streams

Localized corrosion can also occur in areas that experience swirling and turbulence. Protrusions in a pipeline, for example, can cause a slow moving fluid to greatly accelerate the velocity in a small area. The protrusion can be a weld, a weld spatter, a weir, dirt, debris, gravel, or any other deposit on the metal surface. At the protrusion location, corrosion can be greatly accelerated, as shown in Figure 3.35.

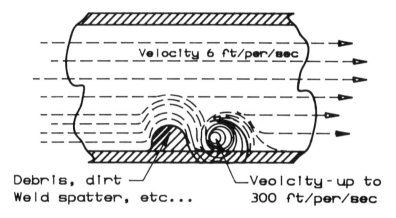

FIGURE 3.35 — *Example of how dirt, weld spatter, etc., can accelerate corrosion in a pipeline.*

Cavitation

There is another manifestation of erosion-corrosion called *cavitation*. This phenomenon has corroded ship propellers for many years. Small bubbles (cavities) in the water suddenly collapse because of a momentary low pressure. This sudden change in pressure causes mechanical damage to the metal because of repeated impact blows produced by the collapse of the bubbles or voids. The designer should specify very smooth surfaces for parts susceptible to cavitation because the number of sites for bubble formation will, therefore, be reduced.

Figure 3.36 displays a piece of a brass pump impeller used for handling process water. Note that the blades are worn thin by cavitation. The pump was replaced by an austenitic stainless steel pump, which has proved to have more resistance to cavitation.

FIGURE 3.36 — *Brass pump impeller used for handling process water; blades were worn thin by cavitation.*

Minimum Velocities

Minimum velocities should also be considered by the designer. To a certain extent, higher velocities reduce the occurrence of concentration cells by making the composition of a solution more uniform over the surface of a metal. For instance, severe pitting will occur on austenitic stainless steel in slow moving or stagnant seawater, while a stainless steel pump impeller handling seawater (with no shutdowns without rinsing with fresh water), will provide good service.

1. An example of the dangers of zero velocity was a *heavy water* plant built in Canada some years ago. Seawater was used as the process fluid because seawater contained a little more heavy water than fresh water. Several types of austenitic stainless steel were used throughout the plant. Seawater was used as a cooling medium as well as for feed stock. Because of technical problems, the plant was shut down for months, with seawater remaining in the equipment. When start-up was attempted, it was discovered that the equipment had pitted so badly that the stainless steel had to be replaced with more resistant alloys at great costs in time and money.

    The lesson here for the designer is to eliminate, as much as possible, all areas in designs that could contribute to stagnant conditions, such as blind flanges, studs, laterals, square corners, undrained by-passes, and any other areas where process fluid can collect and remain in a quiescent condition for long periods of time.

2. Another example of a corrosion problem associated with zero velocity involved a steam turbine that had been used for many years. This turbine was shut down every summer when it was not needed. In the fall of this particular year, when the turbine started up, it did not operate properly because its parts, which were made of several alloys, had been severely corroded. Figure 3.37 shows a number of these parts that were damaged. An investigation disclosed that corrosion had occurred during the months when the turbine had been shut down. No one had drained the condensate from the turbine and that is what caused the excessive corrosion. The turbine blades and the impeller had to be replaced.

FIGURE 3.37 — *Steam turbine parts that corroded because water had not been drained at shutdown.*

What the Designer Can Do to Improve Erosion-Corrosion Resistance

The following are steps a designer can take to improve resistance to erosion-corrosion:

Specify filters. When practical, the designer should incorporate filtering equipment in his design of piping systems that are expected to contain solids, such as slurries.

Avoid intermittent flow. Intermittent flow is worse than steady flow. During the time when the flow has stopped, dirt, debris, sediment, and other suspended solids can settle down in the pipe interior and, depending on the length of down-time, corrosion can occur. When this is expected to happen, the designer should incoporate into the pipeline design, strainers, and other means for conveniently cleaning out the pipeline periodically. Manhole doors, for instance, should be strategically located and, if possible, situated at the lowest points in the system to enable dirt, debris, and other solids to be properly flushed out durinq routine maintenance.

Protect areas with rapid changes in velocity. Locations where there are sudden changes in velocity, as in heat exchangers where the fluid in the bonnet suddenly and greatly increases in velocity as it passes by the tube sheet into the tubes, are susceptible to accelerated corrosion primarily resulting from turbulence. In one case,[25] alloy tubing that handles a process solution was severely corroded in the first 4 in. (102 mm) of the tubing and failed. When the tubes were replaced, they were made 4 in. (102 mm) longer so that they extended beyond the tube sheet into the bonnet. It is reported that the life of the tubes was increased from 2 to 4 years.

Specify ferrules. Ferrules or short readily replaceable pieces of tubing with upset ends, sometimes called *top hats*, can be specified to be the vulnerable ends of the tubes to act as *wear plates*. This is an efficient method. Metal ferrules have proved to be more efficient than plastic or ceramic ferrules.

Heat exhanger tubes with short tubes made of a more turbulence-resisting material welded to the end of the tubing, called *safe-ending*, has also proven to be successful in avoiding the turbulence problem.

Specify reversible equipment. Another method that has been used with attendant savings is the reversible heat exchanger with two bonnets, one at each end. When erosion-corrosion starts to appear at one end of the exchanger, the unit is reversed. In this way, twice the life of the tubes may be attained. Where excessive corrosion is expected, the designer should consider this innovation.

Streamline equipment to reduce entrained air and turbulence. A considerable number of corrosion problems can be avoided in the design stage if the designer makes sure that the fluid is designed to flow smoothly throughout the process with a minimum of turbulence. Bends in piping can be streamlined, and flare- or bell-shaped inlet tubes can be designed to reduce turbulence. For example, at a very large plant where large volumes of fluids are handled at high velocities, it is imperative that the piping be streamlined. Figures 3.38[26] and 3.39[26] show two views of a piping component that will split a process stream without causing unnecessary turbulence. This is an example of excellent design.

This kind of streamlining can also reduce the amount of entrained air. If this is not done, maintenance costs for the completed plant may be high. For instance, inlets to the piping immediately after a pump discharge should be avoided if at all possible. Any equipment or misalignment that will foster turbulence should also be avoided. The use of valves, bends, and branches in the pipe, flanges, flow controllers, etc. should be reduced as much as is practical. Poorly aligned, flanged, or welded joints can also cause turbulence. These cases require tighter installation specifications. Diaphragm valves cause less turbulence than

FIGURE 3.38 — *Streamlined piping component.*[26]

FIGURE 3.39 — *Other view of streamlined piping component.*[26]

gate or globe valves. Obviously, valves that are throttled can cause more turbulence than wide-open valves.

Avoid corrosion in the design stage. Many potential corrosion problems can be avoided by the designer during the design stage, rather than being expensively repaired later by the plant maintenance group after corrosion has reduced the life of equipment. For example, branch pipes set at a 30- to 60-degree angle with the main pipeline cause far less turbulence than right angle joints. [See Figure 3.40(a)]. Figure 3.40(b) shows poor and good designs for a heater tube bundle.[27] The round water box smooths out the flow of the water, while the square water box causes excessive turbulence by reversing the flow. The poor design with the square water box also causes corrosive attack at the periphery caused by cooling water starvation.

Avoid steam flashing. The impingement of steam on metal surfaces can cause local overheating with attendant erosion-corrosion. Wet steam traveling at a high velocity of about 2500 ft/s (45,725 m/min) or more can cause many failures to condenser tubes, turbine blades, etc. The designer can reduce this problem by:

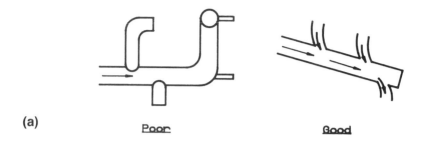

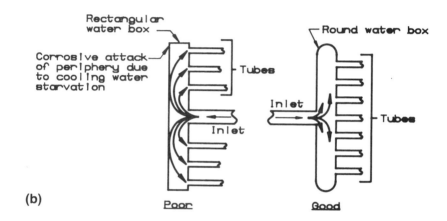

FIGURE 3.40 — *Good and poor designs for fluid flow. (a) Branch piping design; note that the branch pipe with 30- to 60-degree angle with the main piping is a much better design than that with a 90-degree angle. (b) Heater tube bundle design; reversing flow in heater tube bundle is a poor design (left); the round water box (right) smooths out the flow while the square box not only does not smooth out flow, but causes excessive turbulence.*

1. Using baffles to handle the brunt of the attack;
2. Improving the flow design by streamlining;
3. Reducing water droplet content of the steam by better steam control; and by
4. Using a more erosion-corrosion-resistant material.

Alleviate flashing inside heat exchanger tubes. Flashing inside heat exchanger tubes (i.e., the sudden change from liquid to gaseous state) can have a catastrophic result. Figure 3.41 shows a portion of Hastelloy C tubing used for heating phosphoric acid. Erosion-corrosion

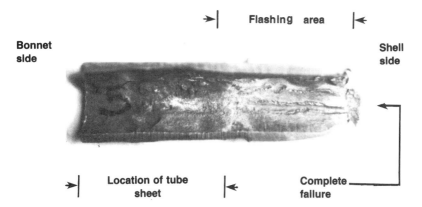

FIGURE 3.41 — *Failure of Hastelloy C tubing from flashing in phosphoric acid service.*

resulting from high velocity flashing grooved the inside diameter of the tube badly and cut through the tubing about 2 1/2 in. (64 mm) away from the tube sheet face. To alleviate this problem, better heating control was implemented.

Select stream velocity properly. The designer should designate the flow velocity of process solutions to be within a range that will minimize corrosion. The highest velocity specified should not exceed that velocity that will cause erosion-corrosion, impingement, or cavitation. The lower limit of velocity should be that velocity which will keep sediment, impurities, etc. suspended in the process solution. The middle point of these two velocity extremes (dependent upon material costs, of course) would probably be the designer's wisest choice.

Specify shutdown procedures. The designer should note in his specifications and especially in his Overall Materials Guide (OMG) (Chapter 5.9) that during the shutdown of process equipment made of steel handling water solutions, the plant maintenance personnel should either completely drain and dry out the equipment, or fill completely with nitrite containing water. This is especially important for equipment containing tubing. Either procedure can avoid pitting. Without adequate shutdown procedures, pitting during shutdown can be disastrous, depending on the length of the shutdown.

☐ 3.5 Fretting Corrosion

Fretting Corrosion Mechanism

Fretting corrosion occurs at the interface between two highly loaded, close-fitting surfaces when these surfaces are subject to slight relative motion causing slip, galling, and chaffing. Rapid attack will occur when this action is accomplished in the presence of oxygen. This type of corrosion has been found in bolted flanges, couplings, shafts, connecting rods, brackets, rivets, washers, etc.

There are two fundamental actions in fretting corrosion, physical and corrosive effects. It is thought that the primary action is a physical one based on the molecular attraction between the two surfaces positioned tightly against one another. Cycles of seizure and pulling away produce tiny particles that are heated because of the friction caused when they are torn loose. The corrosive action is when oxygen reacts with the particle to form an oxide corrosion product.

Austenitic stainless steels (300 Series), and also titanium and aluminum, are the most susceptible to this form of corrosion since they are relatively soft and show more tendency to seize. As mentioned in previous chapters, these metals have a tightly adherent oxide film that protects them from corrosion. Apparently, the rubbing and seizing during fretting produces a plucking action on the surface that removes this protective film and thus allows fretting corrosion to occur.

Example of Fretting Corrosion

A classic case of fretting corrosion happened years ago and caused much concern for the engineers involved. New automobiles were being shipped by rail to distributors. The cars were supported on their wheels. When the cars reached their destination, much to everyone's surprise, all the wheel bearings had to be replaced because they had corroded. Even though the automobiles had been tightly strapped down and the wheels chocked, there was still enough relative motion (resulting from the vibration of the train) on the bearings supporting the weight of the car to cause fretting corrosion. The problem was lessened by supporting subsequent shipments of cars by their axles instead of the wheels, thus eliminating the high stress on the bearings.

What the Designer Can Do to Prevent Fretting Corrosion

When designing close-fitting parts that will bear a substantial load under vibratory conditions, the designer should heed the following guidelines:

1. If at all possible, design the parts to exhibit less or no relative motion. (This may be difficult to do.)
2. Avoid high finishes on parts because the higher the finish, the greater will be the fretting corrosion.

3. Avoid specifying that the finish on the two mating surfaces be the same since this will aggravate corrosion. Always design the two surfaces to have a marked difference in finish.
4. Using grit-blasting on one surface can be beneficial since it tends to reduce or stop relative motion of the parts by a gripping effect.
5. Using nickel- or chromium-plated surfaces will reduce this potential of fretting corrosion, especially under wet conditions.
6. Specify molybdenum sulfide to act as a lubricant. Lubricants help reduce friction and seal out oxygen. For some reason, petroleum-base lubricants do not work as well as molybdenum sulfide.
7. For dry conditions, the best metal to specify for one of the surfaces is brass. Austenitic stainless steel should not be specified.
8. When the design requires excellent corrosion resistance where austenitic stainless would normally be specified, except for its poor fretting corrosion property, specify a martensitic stainless steel that can be hardened, such as AISI 440, 420, or 416.
9. Specify the use of cements to seal out the air.
10. When feasible, specify that one of the surfaces be a nonmetallic material with a low coefficient of friction, such as Teflon.

☐ 3.6 Pitting

Pitting Mechanism

As the name implies, *pitting corrosion* proceeds in a very localized area with the surrounding area relatively untouched. The corrosion may proceed at a very rapid rate. In stagnant solutions, in recesses, outside of recesses, and under deposits are the most likely areas where pitting will occur. Sometimes, this form of corrosion is difficult to detect, and it can cause disastrous results. The most likely candidates for pitting are metals protected by thin films such as copper, stainless steel, aluminum, titanium, and magnesium. It may also occur in steel, iron, lead, and many other metals. Pitting probably accounts for the greatest number of unexpected corrosion failures.

1. Pits may or may not fill up with corrosion products. When they do, such products can cause a cap to form over the pit cavities and are then called *tubercles*, *mounds*, or *nodules*.

FIGURE 3.42 — *Example of how pits form under water drops.*

When a pitting environment is suspected, the wall thickness of parts should be increased when practical. Obviously, the thinner the section, the greater the pitting hazard will be.

2. A drop of water on a steel plate can cause pitting, as depicted in Figure 3.42. The area under the water in the center is shielded from the air and thus becomes the anode resulting from a low concentration of oxygen where the pit or pits are formed. Just inside the outer periphery of the water drop, the area is more available to oxygen and thus is higher in oxygen concentration and consequently becomes the cathode.
3. Dirt, sediment, debris, etc. in contact with metals in the presence of an electrolyte can also cause pitting. Marine growths, such as barnacles which adhere to metals in seawater, likewise can cause corrosion cells to be set up and result in pitting.
4. Foreign inclusions in metals, such as slag and sulfides in steel, can also trigger pitting, as can any heterogeneity of the metal itself.
5. Dead spaces, such as blind flanges can foster pitting by permitting dirt and sediment to collect.

Examples of Pitting

Below are 4 field examples of pitting.

Example 1 — AISI 304 stainless steel 16-gallon (61-liter) drums were used to ship 65% nitric acid from a plant in the midwest. The top heads of the drums were flat to make stacking easy. Calls were received to come to the plant because "the acid was corroding the stainless steel drums causing some of them to leak." An investigation disclosed that the flat tops had many pinholes, and a few that had progressed all the way through the wall from the outside to the inside, not the other way around. The acid, as suspected, was not the corrosive agent. It was gravel inadvertently lying on the top of the drum plus water that collected on the top of the drum containing chlorides from the plant atmosphere that caused severe pitting underneath the gravel.

Pitting can occur when certain materials like chlorides on austenitic stainless steel remove small areas of the passive film on the surface of a metal. This causes an active-passive cell to be set up; the anode or active area is the small area where the protective film has been removed from the metal. The passive area or cathode is the surrounding area still possessing its protective film. The corrosion current density is large because there is a tiny anode and a comparatively large cathode. Under these conditions, corrosion penetration can proceed rapidly. Figure 3.43 shows a photomicrograph at 5X magnification of a pit found in one of these drums. This pit was just starting to penetrate the top head. Note that the external diameter of the pit is only about half the diameter inside the pit.

FIGURE 3.43 — *Pit at 5X magnification in top head of an AISI 316 stainless steel, 16-gallon (61-liter) nitric acid drum.*

Pitting attack is dangerous because failure can occur in metals without any appreciable weight loss.

Example 2 — An example of this form of attack was observed during an inspection of a large steel containment vessel that had thermal insulation covering its exterior. The insulation looked intact and in good condition. A pencil pushed through the insulation at the bottom of the vessel released a stream of water as though a faucet has been turned on. Rain had seeped behind the insulation. A section of the insulation was removed and heavy pitting was observed on the steel wall. If the original design had called for some means of keeping the water out of the insulation, such as a light aluminum water-tight jacket or plastic covering, the pitting damage would not have occurred.

Example 3 — Figure 3.44 shows a badly pitted copper tube from an air conditioning cooling system in a large hotel. Because the water flow had diminished as a result of the sediment which had collected in the copper tubes, it had been decided to clean out the tubes with sulfamic acid. The acid not only did an excellent cleaning job but also opened up a lot of pits that had been in the tubes previously under the sediment. The pits had remained clogged up until the cleaning operation. The cleaning should have been specified at regular intervals rather than waiting until pitting had destroyed the thin-walled tubing. Note that the pits are all in a row in the bottom of the tube where the sediment collected.

FIGURE 3.44 — *Pitting in copper tube from air conditioning system.*

What the Designer Can Do to Reduce Pitting

The designer can take the following actions to reduce pitting:

1. Make sure that any corrosion rate data to be used for selecting materials of construction represents general corrosion attack. If there is a question about the corrosion profile, ask for pitting factors to be furnished or determined. (Pitting factors are the ratio of the maximum measured pit depth to the average penetration calculated from weight loss.)
2. Since pitting is electrochemical in nature, specify cathodic protection to stop this form of corrosion. The use of a zinc-rich coating on steel or aluminum can accomplish this at a low cost.
3. Do not specify insulation, wood, cloth, felt, or other substances that will absorb or retain water for applications where the metal surface becomes wet periodically. This can result in pitting. This is sometimes termed *poultice corrosion*.
4. In instructions to plant maintenance, specify that periodic cleaning must be done for specific equipment and piping where there is a pitting potential.
5. In designs, as much as possible, avoid using areas such as pockets, sharp corners, re-entrant angles, dead legs, or other geometry that can cause fluids or solids to collect.
6. Avoid designs where water spray is not confined, since this can cause pitting of steel surfaces. Specify baffles and better spray nozzles.
7. Specify that during shutdown, all equipment be drained of solution or that arrangements be made to keep the solution moving over the metal surfaces. (Refer to Chapter 4.4.)
8. Specify that a protective coating be applied to the metal surfaces.
9. Increase the thickness of the parts where pitting is likely to occur.

☐ 3.7 Microbiologically Influenced Corrosion

Definition of Microbiologically Influenced Corrosion

Microbiologically influenced corrosion (MIC) is defined as corrosion that is initiated or accelerated by micro-organisms. Corrosion would be nonexistent or much less severe in the absence of such organisms. The manifestations of MIC are usually deposits in the form of mounds and tubercules on the surfaces of process equipment. Certain forms of MIC also develop into irregular buildups. Underneath these deposits, evidence of corrosive attack on metals and alloys can be found. MIC has been recognized in the petroleum industry and municipal sewer systems for many years. However, only in the last decade or so has the chemical industry recognized MIC as a major corrosion problem. There is still controversy over the extent of MIC.

Bacteria Causing MIC

As explained by Tatnall,[28,29] bacteria are simple one-celled organisms that form colonies, assimilate nutrients from their surroundings, derive energy for life and procreation by oxidizing these nutrients, and excrete the byproducts of the oxidization process. This causes changes in the environment in local areas, which in turn, causes MIC.

There are innumerable species of bacteria in existence. Many cause no corrosion damage at all; however, there are a number of species that do cause MIC. A few of them are:

— *Sulfate reducers* that use the sulfate ion as an oxidation agent for the assimilation of organic matter;[30]
— *Sulfur oxidizers* that oxidize sulfides to sulfate, thus producing sulfuric acid. They also have the ability to oxidize dissolved ferrous iron to insoluble ferric hydrate. Certain varieties will oxidize and concentrate manganese. These bacteria are aerobic and create oxygen depletion under tubercules; and
— *Gallionella bacteria* that produces deposits containing iron and manganese. They also have the ability to concentrate chlorides, which creates ferric and manganic chlorides that are acidic. These chlorides can cause general corrosion in steel and aggressive pitting of austenitic stainless steel under ambient temperatures and stress corrosion cracking under heated conditions.

Examples of MIC

A series of photos by G. Kobrin shows how gallionella bacteria attacks austenitic stainless steel welds. Figure 3.45 shows typical mounds along a weld seam. Figure 3.46 is a closeup of a single mound. Figure 3.47 displays a mound with the cap broken away. Figure 3.48 shows an area in Figure 3.47 after cleaning with wire brush; note the apparent absence of corrosion. Figure 3.49 exhibits the area of Figure 3.48 after subsurface cavity had been partially opened with a pick. Figure 3.50 displays a cross section of stainless steel weld showing subsurface cavity that was formed under the gallionella mound. Figure 3.51 shows a sidewall weld in a stainless steel tank after probing open subsurface cavities, and Figure 3.52 shows external weeping in a

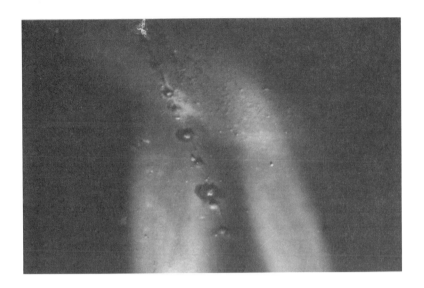

FIGURE 3.45 — *Gallionella mounds along a weld seam floor of a stainless steel tank.*

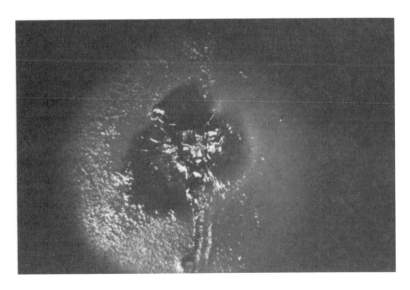

FIGURE 3.46 — *Closeup of a single mound showing a slimy appearance when fresh and wet.*

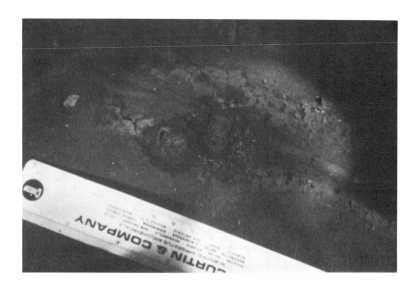

FIGURE 3.47 — Dried mound similar to Figure 3.46 only with the cap broken away.

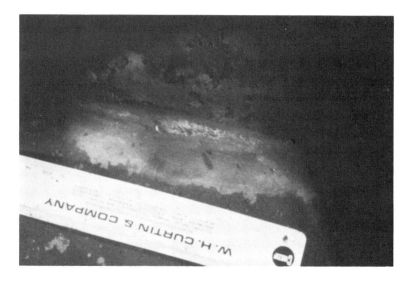

FIGURE 3.48 — Area in Figure 3.47 after cleaning with a wire brush; note the apparent absence of corrosion.

FIGURE 3.49 — Area of Figure 3.48 after subsurface cavity was partially opened up with a pick.

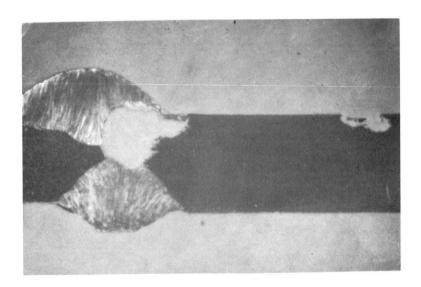

FIGURE 3.50 — Cross section of stainless steel weld showing subsurface cavity under gallionella mound.

FIGURE 3.51 — Side wall weld in a stainless steel tank after prying open the cavities; note the typical vertical streaks.

FIGURE 3.52 — External weeping along a stainless steel weld seam under gallionella mounds in a recirculating cooling water system.

recirculating cooling water system resulting from attack under gallionella mounds.

What the Designer Can Do to Reduce MIC

When a plant is being designed for a location where the available water supply has been known to cause MIC, there are several remedies that the designer can use. (If the water supply has not been characterized, it should, of course, be tested.) The following preventive steps can be taken:

Change materials. The materials of construction can be changed to alloys that are resistant to MIC. A designer's first reaction may be to change from using carbon steel equipment to using stainless steel equipment. If the gallionella bacteria is present in the water supply, the stainless steel may also be vulnerable to MIC. The use of monolithic reinforced plastic equipment or steel protected by thick film linings often can avoid the MIC problem.

For example, Kobrin[31] reports that reinforced epoxy coatings have prevented MIC in a waste water treatment system made of carbon steel and AISI 304 stainless steel. Prior to the application of this thick film lining, the system had been aggressively corroded, leaving deep local attack on the carbon steel and pits in the stainless steel.

The designer should specify appropriate coatings initially when the plant is being designed. By recognizing the MIC problem and selecting the correct lining to be installed, the high expense of installation in the field would be avoided. Because surface preparation is very important to the lining's integrity, a lining installed in the field over a surface that may be difficult to clean and is, of course, not as dependable as one installed in the vendor's shop.

Change the water supply. When there is a choice of water source, such as from wells, rivers, and municipal water works, the bacteria-free source should, of course, be selected. This alternative would certainly be the least expensive; however, plants seldom have such a choice.

Control the bacteria. Controlling bacteria is probably the most expensive alternative; however, in some cases, it is the only option. When controlling bacteria is contemplated, it is best to confer with water treatment experts who have years of experience in this field. The use of commercial biocides requires a lot of testing because there is such a wide variety of biocides available and the various harmful colonies of bacteria require different biocides for their extinction or control. The various virulent strains must be identified through tests and then the optimum biocides for the water involved must be thoroughly tested.

Afterwards, methods must be worked out to dispense the chemicals and to determine the quantities that will be required. It is best to call the water treating consultants early so that these tests can be run on the field water supply before the plant is built. The designer should insist that this is done.

Avoid hydrostatic testing. Surprisingly, numerous cases of corroded equipment resulting from MIC has been reported to be caused by the use of contaminated water during hydrostatic testing. Even though the water is emptied from the equipment after testing, bacteria colonies may remain in the equipment. During subsequent use, these bacteria colonies may thrive and develop in the new environment. Infestation of these colonies can subsequently cause corrosion attack that may not show up until much time has elapsed. Several things can be done to assure that the hydrostatic testing does not foster corrosion problems, such as:

1. Completely avoid using contaminated water. At one plant where many pieces of equipment had to be hydrostatically tested, a batch of water for testing was prepared and determined to contain no harmful bacteria. Then the water was used over and over again. Uncontaminated water can be produced by pasteurizing water at a temperature of 160 F (71 C) or higher.
2. If only one source of water is available for hydrostatic testing, such as a river, leave the water in the vessel for as short a time as possible and then thoroughly clean and dry the equipment. It takes a good amount of time for MIC to develop. In one classic case, to illustrate this point, river water was put into a number of empty stainless steel tanks to act as ballast for an impending hurricane. After the storm had passed, the contaminated water was inadvertently left in the tank for many months. Then it was discovered that aggressive pitting resulting from MIC had occurred, making it necessary to repair the tanks by extensive welding.
3. If steam condensate is available, it may be used for hydrotesting.
4. Treating contaminated water with biocides to destroy bacteria (particularly gallionella when austenitic stainless steel equipment is involved) is a safe procedure.

Note: It should not be construed that all water will cause MIC during hydrotesting of tanks and equipment. As a matter of fact, it happens very infrequently. However, when it does happen, it can be disastrous; therefore, the designer should insist on knowing the quality of the water prior to hydrostatic testing. This request should be part of his specification.

☐ 3.8 Stress Corrosion Cracking

Stress Corrosion Cracking Mechanism

Stress corrosion cracking (SCC) has been described as the spontaneous brittle fracture of a susceptible material (usually itself quite ductile) under tensile stress in a specific environment over a period of time. Most commercial metals will crack in particular environments and under certain conditions; however, there is not one environment that will crack all metals. Each of these specific environments acting alone would cause little corrosion on a metal when stresses are low. It is the combination of the right level of stress and a particular environment that will produce SCC. Such stresses may be a combination of applied and internal stresses or internal stresses alone. Cracking is either intergranular (between grains) or transgranular (across grains) or both.

The mechanism that fosters SCC appears to be electrochemical and mechanical in nature. However, the basic cause is not fully known. Although SCC has been investigated over a period of years, these studies have primarily covered the characteristics of such corrosion. Despite both laboratory and industrial experimentation, the exact mechanism of SCC remains obscure. There have been many different theories advanced to explain the mechanism. For further review, see References 32 and 33.

Specific Environments That Can Cause Stress Corrosion Cracking

Although in recent years austenitic stainless steels have received the most amount of attention regarding SCC failures, practically no metal is immune to this type of attack. Table 3.4 lists some of the environments that cause SCC in some metals. In each case listed, there must be special conditions, such as concentration and temperature, that will trigger the reaction. SCC is insidious because there is usually no manifestation of trouble before cracking appears. Failures of this type can be unexpected and hence hazardous and expensive. It is a major problem in the chemical and petroleum industries.

Cracking Vulnerability of Various Metals in Diverse Environments

Carbon steels. Many environments can cause carbon steels to fail

TABLE 3.4
Alloy-Environmental Systems Susceptible to Stress Corrosion Cracking

Alloy	Environment
Carbon Steel	Carbonates and Bicarbonates (reversion to caustic)
	Caustic
	Nitrate
	Hydrogen (attack)
	Cyanide
	Anhydrous Ammonia
	$CO/CO_2/H_2O$ Mixtures
Austenitic Stainless Steel	Organic and Inorganic Chlorides
	Acidic Hydrogen Sulfide
	Caustic
	Sulfurous and Polythionic Acids
	Nitrates
	Hydroxide
	Chlorides
Nickel-Base Alloys	Caustic above 600 F (above 50%) (Alloys 200, 400, and 600)
	Mercurous Nitrate (Alloy 400)
	Mercury (Alloy 400)
	Hydrofluoric Acid and Oxygen (Alloy 400)
	Fluosilicic Acid (Alloy 400)
Copper-Base Alloys	Amines (breaks down to ammonia)
	Dilute Ammonia
	Ammonium Hydroxide
	Mercury
	Sulfur Dioxide
Aluminum Alloys	Seawater
Titanium	Organic Chlorides above 550 F
	Hydrogen Embrittlement
	Methanol
	Seawater (sodium chloride)
	Nitrogen-Tetroxide
Tantalum	Hydrogen Embrittlement

by SCC. Among these environments are caustic, nitrates, amines, ammonia, cyanides, hydrofluoric acid, and sulfides. Cracking is mostly intergranular and is related to the composition of the steel and whether the steel has been heat treated or not. As a general rule, steel will not stress crack in any of these environments if it has been properly stress relieved and unusually high stresses are not imposed.

When steel fails under caustic conditions, it is called *caustic embrittlement* or *boiler embrittlement*. Figure 3.53 shows a typical stress crack in the wall of a boiler. Figure 3.54 shows an area of AISI 1020 steel at 100X magnification displaying cracking in a hot caustic environment. NACE T-5A Committee has studied SCC of steel in caustic environments by reviewing hundreds of cases of steel used throughout industry. The results of this study are summarized in Figure 3.55, which relates temperatures and percent concentration in incidents of failure. No cases of SCC were found below the line on the graph.

Copper-base alloys. Copper-base alloys are susceptible to a

FIGURE 3.53 — *Caustic embrittlement in a boiler. (Note the typical stress crack.)*

variety of environments that will cause SCC, including amines, nitrates, ammonia, and mercury. *Season cracking* is the term generally applied to the cracking of copper-base alloys. In one example, admiralty brass tubes failed rapidly in a heat exchanger by SCC. An analysis of the process gas stream being handled by the exchanger showed the unsuspected presence of ammonia that caused the cracking. Figure

FIGURE 3.54 — *Cracked area of AISI 1020 steel at 100X magnification. (The environment was hot caustic.)*

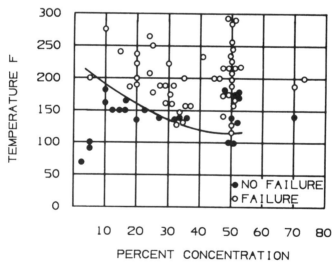

FIGURE 3.55 — *Graph compiled by NACE T-5A Committee relating temperature and concentration of caustic to incidents of failures and no failures by SCC.*

3.56 displays one of the cracked tubes. The tubes were replaced with bimetallic tubes.

An unusual example[26] of SCC was the failure of a static ground wire at a very large plant in the south. Such a wire is actually a grounded wire that is suspended above a power line as a lightning arrester. The three strands composing the wire were 0.078 in. (1.981 mm) in diameter and made of bronze (95.5% Cu, 2% Sn, 2.5% Al). The wire had been in service for 23 years before this final failure occurred.

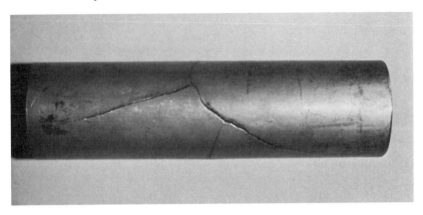

FIGURE 3.56 —*Admiralty tube that cracked because of ammonia.*

Figure 3.57 shows the fracture area at 5X of one of the strands that failed. Numerous cracks were discovered along the length of the static wire, as shown in Figure 3.58. Sections of the fractured and cracked wire at 5X magnification are shown in Figure 3.59. It was concluded that SCC was the cause of the failure. The atmosphere was the corrosive environment while tensile and bending stresses were applied to the wire over the years.

An investigation of an aluminum-coated steel wire phase conductor and a galvanized steel static wire that had also been in service for 23 years showed no significant corrosion. Based on this, all the bronze static wire was replaced with aluminum-coated steel containing about 25% aluminum in the cross-sectional area. Figure 3.60 exhibits longitudinal sections of three-strand *Alumoweld*[†] static wire ground line that was installed. It is interesting to note that a steel wire cathodically protected with aluminum performed much better over a long period of time than a bronze wire.

[†]Registered trade name.

FIGURE 3.57 — *Failure of bronze static wire ground line (5X magnification).*

FIGURE 3.58 — *Typical cracks in bronze static wire.*

Pure copper is not susceptible to SCC, however, the addition of less than 0.2% phosphorus, arsenic, or antimony makes copper prone to cracking in ammoniacal atmospheres. As recorded in history, Napoleon's army had trouble with brass shells cracking and splitting. It turned out that the shells were stored in the stables with the horses. The ammonia from the horses' urine triggered SCC of the cold-worked brass shells.

FIGURE 3.59 — *Sections of bronze static wire showing cracks.*

Copper alloys are very sensitive to mercury. Many failures have occurred this way in industry. As an example,[16] during a plant startup, mercury manometers were used to check the pressure drop on the waterside of condensers. Because of the corrosive nature of the plant water, bimetallic tubes were used with admiralty metal on the water side and steel on the process side. Inadvertently, mercury was sucked from the manometers. After one month, severe cracking occurred on the admiralty metal and three condensers had to be retubed.

In another instance, an admiralty brass air compressor cooler failed shortly after startup. An examination disclosed that the tubes that had failed showed season cracking, typical of ammonia. Further investigation disclosed that the power house where the compressor was located was about 100 yards (91 m) away from an operating building where chemically pure ammonium hydroxide was processed. The ammonium fumes that had caused the cracking could not be detected by smell from the power house.

FIGURE 3.60 — *Metallurgical mounts show longitudinal sections of three-strand Alumoweld static wire that was installed. Light areas are aluminum cladding and dark areas are steel core.*

Austenitic stainless steel. Monack analyzed the incidents of SCC of steel, copper alloys, stainless steel, high-nickel alloys, titanium, and tantalum reported in the Du Pont study covered in Chapter 1.5. He found that out of 89 SCC cases reported, 57 or 64% occurred to austenitic stainless steel by chlorides over a four-year period, while season cracking of copper-base alloys comprised the second highest percentage at a distant 8%. (See Table 3.5.) Austenitic stainless steels are the work horses of the chemical industry, and through experience, it has become apparent that SCC is a major weakness in austenitic stainless steels, a fact a designer must always consider when planning or specifying the use of these alloys. SCC is caused when the following three conditions are satisfied:

1. Stainless steel is in the wrong environment (chloride environments are the worst);
2. Temperature is above 60 C (140 F); and
3. Stresses are present. (Even if the fabricated stainless steel has been completely annealed, there are still high enough residual stresses to cause SCC.)

TABLE 3.5
Frequency of Stress Corrosion Cracking Failures[1]

	No. of Failures	% of Failures
Steel		
Cyanide	1	1
Caustic	4	5
Hydrogen Attack	4	5
Nitrate	4	5
Copper Alloys		
Amines	1	1
Carbon Dioxide	1	1
Season Cracking	7	8
Stainless Steel		
Acidic Hydrogen Sulfide	1	1
Chlorides	57	64
High Nickel Alloys		
HCl	1	1
HF	4	5
Caustic	1	1
Organic Chlorides	1	1
Titanium		
Hydrogen Embrittlement	1	1
Tantalum		
Hydrogen Embrittlement	1	1

[1] Breakdown of 89 cases of SCC over a 4-year period; based on Monack's study.[10]

Failures are frequently found on the cooling water side of condensers resulting from the concentration of chlorides in the water. Figure 3.61 shows a photomicrograph at 150X magnification taken from a stainless steel superheater tube that had cracked during startup. This was caused by the accidental contamination with a cleaning agent containing chlorides that was used to clean the boiler prior to the startup of a plant. Most of the superheater tubes had to be replaced at a high cost.

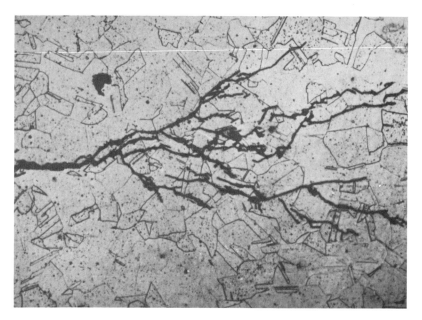

FIGURE 3.61 — *Typical austenitic stainless steel inter- and transgranular— many branched cracks found in failed superheater tubes. (Original magnification was 150X.)*

The vulnerability of austenitic stainless steels to the chloride ion when stress and a high enough temperature are involved has been emphasized before. Although there are frequent cases in which cracking does occur, there are also numerous cases where austenitic stainless steels have been used in the presence of chlorides with no cracking. Many applications show that austenitic stainless steels are not cracked in solutions containing even high concentrations of chlorides if the temperature is ambient. Therefore, the designer should not rule out the use of the very useful austenitic stainless steels until a careful study is made.

Nickel-base alloys. Nickel-base alloys are generally considered to be immune to SCC; however, there are a few exceptions, such as Alloy 625 (Inconel†) cracking in hot caustic service over 315 C (600 F) and Alloy 600 (Monel†) cracking when exposed to wet vapors of hydrofluoric acid containing oxygen when the alloy is in a highly stressed condition.

Lackey[34] reports an unusual example of the SCC of Monel. Flexible AISI 304 stainless steel steam hose was used for alternating wet and

---

†Registered trade name.

dry cycles. It promptly failed because of the presence of chlorides. The stainless steel hose was replaced with Monel hose, which appeared to be the obvious answer because it is supposed to be resistant to chloride SCC. The Monel failed by SCC almost as quickly as the stainless steel had. (See Figure 3.62.) The condensate was analyzed and found to contain amines that were thought to have caused the failure of the highly stressed Monel Alloy 400 hose. Flexible tubing made of Alloy 800 (Incoloy) replaced the Alloy 400 hose, which has shown satisfactory service without cracking.

FIGURE 3.62 — *Stress corrosion cracking of Monel 400 corrugated hose.*

Titanium. Titanium and its alloys have excellent corrosion resistance in many media. However, these metals are sometimes susceptible to SCC in a few environments. In organic chloride solutions at above 290 C (550 F) for instance, this type of corrosion will occur. Methyl alcohol and commercial nitrogen tetroxide N_2O_4 at ambient temperatures have both caused cracking of titanium alloy tanks. As an example, titanium alloy tanks for one of the first Apollo space missions cracked at room temperature when they were washed out with methanol. Titanium parts used in high-temperature service have cracked because of the presence of sodium chlorides and other chloride salts.

What the Designer Can Do to Reduce Stress Corrosion Cracking

The following are rules of thumb a designer can use to reduce SCC:

Select the best material of construction. Tables 3.4 and 3.5 can be used as guides for the designer as to what combinations of alloys and environments to avoid. Some of the alloys that are known to resist SCC (remembering the limitation of Alloy 600 stated before) are:

> Incoloy Alloy 800
> Inconel Alloy 600
> Inconel Alloy 625
> Hastelloy C-276
> Hastelloy G-3
> Carpenter 20 Cb-3

Generally, alloys containing more than about 29% nickel are not susceptible to SCC; however, there are some exceptions, as noted in Table 3.4. In some cases, bimetal tubes are used to combat SCC. For instance, tubes with 90-10 cupro-nickel on the water side and an appropriate alloy on the process side is often specified.

Table 3.6[35] shows SCC data of various alloys compared to AISI 304 and 316L stainless steels in boiling 42% magnesium chloride. This solution is a powerful crack former for most alloys and is consequently used as a comparison medium. The disadvantage of selecting SCC-resistant alloys is that they are generally more expensive than austenitic stainless steels.

Lower the internal stresses. The normally applied stresses are usually of such a low magnitude that they will not materially affect SCC

TABLE 3.6[35]
Comparative Stress Corrosion Cracking Data

Alloy	Nominal Nickel Content (%)	Time to Crack in Boiling 42% Magnesium Chloride (hours)
AISI 304 Stainless Steel	10	1 to 2
AISI 316L Stainless Steel	12	1 to 2
Alloy 825	32	46
Alloy 625	72	no cracks in 1000
Hastelloy G	42	no cracks in 1000
Hastelloy C-276	59	no cracks in 1000

tendencies. However, locked-in stresses resulting from welding and localized stresses caused by stress raisers in a structure are very instrumental in inducing this type of corrosion. The designer should pinpoint areas where there are stress raisers in his designs and eliminate them when practical. Cold working of austenitic stainless steel also creates extremely high stresses. A good example of how cold working can promote SCC is displayed in Figure 3.63. One end of an annealed AISI 304 stainless steel tube was severely cold worked while the other end was left in the annealed condition. The tube was then exposed to a heated chloride solution. Note that the cold-worked end cracked badly while the other end was relatively unaffected.

Annealed End

Cold-worked End

FIGURE 3.63 — *Just one end of annealed AISI 304 stainless steel tubing was severely cold worked and then the tubing was exposed to a heated chloride solution. (Note the severe cracking on the cold-worked end.)*

The designer should remember that even if a tubing is completely annealed and then bent, the bent areas can be prone to SCC. Also, when straightening operations are performed on annealed tubing, high stress areas can be created. Punching, cutting, machining, shearing, crimping, or any other operation that requires cold work can also build up high stresses. (See Figure 3.64.) In this figure, the bubble cap shown is made of AISI 410 stainless steel. The reason for the failure was that the bubble cap had been deep drawn and punched with no stress relief. A great number of these bubble caps failed within the first two weeks after startup. When stress-relieved caps of the same alloy were used, no cracking occurred.

Welds (particularly fillet welds) are perhaps the greatest contributors of high locked-in tensile stresses. This is because of the innate

FIGURE 3.64 — *Bubble cap made of AISI 410 stainless steel cracked because no stress relief was given after deep drawing and punching.*

property of deposited liquid metal to shrink when it solidifies. Such stresses often approach the yield point of the metal. An example of this was the local cracking on the walls of a steel waste storage tank handling a caustic solution at a temperature of approximately 104 C (220 F). Cracks were formed around fillet welds used to attach brackets and lifting lugs to the inside wall of the vessel. These welds created extremely high localized stresses which, in conjunction with the caustic environment and the high temperature, caused SCC. When subsequent tanks were built, they were completely stress relieved and as a consequence, no further corrosion cracking problems were experienced.

Moller[36] reports on a failure of AISI 304H stainless steel tubing in a superheater. The tubes were cracked at high stress areas in heavy fillet welds joining the spacer lugs to the outside of the heavy-walled tubing. The cracks originated at the top of welds, as had occurred with the fillet welds in the waste storage tanks mentioned previously. A design change was made in the superheater so that the spacers were bolted to the tubes instead of being welded. This eliminated the high stresses and consequent failures.

The designer can also avoid locked-up stresses by not including shrink fits, press fits, and tapered pins in his designs where SCC may be prevalent. The use of fittings that have to be forced into place also create high levels of stress. Lack of precise specifications can allow the purchase of such fittings.

Occasionally, austenitic stainless steel equipment or parts are annealed or stress relieved to reduce stresses. However, although such treatments do help, SCC is not necessarily avoided. The reason is that even after the stresses have been "relieved," there are internal stresses still present high enough to cause SCC. These stresses result from unequal cooling from the stress-relieving temperature and internal volume changes caused by the rearrangement of the microstructure. Since there are no guarantees that stresses are sufficiently relieved by heat treatment, the designer should not rely on stress relieving or annealing alone to eliminate SCC.

Use lower temperatures. The temperature and the concentration of the corrodent are both important in causing SCC. A rule of thumb used in deciding whether a solution containing chlorides would crack austenitic stainless steels is if the temperature is 60 C (140 F) or below, no cracking will occur. For all alloys susceptible to SCC, the lowering of the process temperatures can reduce and many times eliminate any attack.

Fishing trawlers operating in the North Atlantic Ocean, where the temperature of the seawater is generally low, have been using austenitic stainless steel gear successfully for many years with no reported SCC incidences. The cracking of stainless steel heat exchangers seldom occurs at temperatures below 60 C (140 F).

The designer should strive to keep the temperature as low as possible. As an example, suppose the process requires hot gases at 125 C (257 F) to be condensed in AISI 304 stainless steel condenser tubes in a heat exchanger by plant water that is available at 35 C (95 F). (See Figure 3.65.) The conventional design would be to introduce the water at bottom to flow counter-current to the top. Since the hottest gases contact the hottest water, the water exits at 85 C (185 F), thereby creating a temperature zone at the top of the heat exchanger conducive to cracking of the austenitic stainless steel tubes. Simply by significantly increasing the flow of water, the designer can lower the temperature. However, this can be expensive because it requires more water and possibly larger diameter tubes. For the designer, the most efficient way to lower temperature would be to change the counter-current flow to co-current flow so that the coolest water contacts the hottest gas at the top. The only requirements are a throttling valve on the water exit to assure that the heat exchanger always remains full of water and perhaps a little larger heat transfer area. Since the temperature at the top is now lowered to 55 C (131 F), there is little danger the tubes will crack.

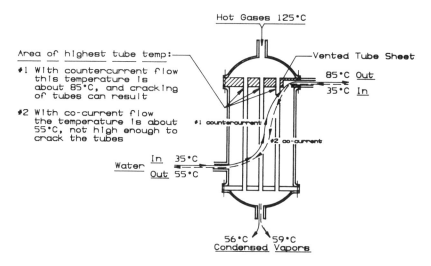

FIGURE 3.65 — *Reversing and increasing cooling water flow reduces maximum temperature of tubes.*

It is true that there have been very few reports of SCC failures in austenitic stainless steels at temperatures below 60 C (140 F). However, tests conducted using the Dana and DeLong wick test were able to crack specimens of stainless steel at 40 C (104 F). This is about the lowest temperature reported that has caused cracking. On the other hand, tests conducted on submerged specimens, where chlorides could not concentrate, did not crack at a temperature of 50 C (120 F), even though the chloride content was 20,000 parts per million (ppm).

Reduce chloride concentration. The rule of thumb used by many engineers dealing with SCC of austenitic stainless steels in chloride-containing solutions is to keep the temperature below 60 C (140 F), as just discussed, and to hold the chloride concentration at below 100 ppm. Like many rules of thumb, this one is very misleading because actually austenitic stainless steels can resist SCC in much higher concentrations of chlorides. It is generally thought that completely immersed austenitic stainless steel in neutral or alkaline solutions containing high chloride content will not crack. The corrosion problem arises when the area being exposed is a heat transfer surface on which localized boiling can occur that will concentrate the chlorides. The designer should remember that if there are any chlorides present, they can eventually concentrate to strengths that can crack austenitic stainless steels. For instance, we have observed SCC of AISI 304L stainless steel in reactors when there was but a few parts per billion (ppb) of chlorides present in the process solution. The cracking damage had occurred in crevices where the chlorides had managed to concentrate to several percent.

Therefore, the designer should be most concerned with the ease with which his designs can concentrate chlorides rather than with the original concentration of chlorides. A case in point is the design of vertical condensers. The left side of Figure 3.66 shows the usual or standard design that has a vapor space between the top tube sheet and

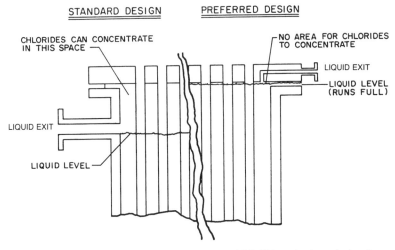

FIGURE 3.66 — *Vented tube sheet (right side). This is the best design for eliminating chloride stress corrosion cracking.*

the liquid level. A series of failures of a number of AISI 316 stainless steel condensers of this design will illustrate the problem. Severe cracking was observed after 6 to 12 months of service near the upper ends of the tubes, as shown in Figure 3.67. Hot process gases entered

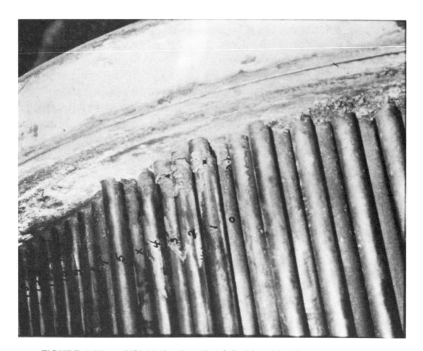

FIGURE 3.67 — *AISI 304L tubes that failed by chloride stress corrosion cracking in air space above tubes. (Note the deposit of salt in the upper right-hand corner.)*

the top of these units at 155 C (311 F) and condensed liquors left the bottom at 60 C (140 F). Fresh water containing 100 ppm of chlorides was used for cooling in counter-current flow entering the bottom at 35 C (95 F) and leaving at 80 C (176 F). Inspection disclosed a buildup of salt deposits on the tubes in the area where cracking occurred. The vertical condensers were redesigned, as shown on the right side of Figure 3.66. By drilling the top tube sheet as shown, the vapor space was vented and the cooling water flowed out through this unit, leaving the tubes completely immersed in the cooling water. (This is also shown in Figure 3.65.) This action eliminated cracking in the tubes.

Another solution to the SCC problems of heat exchangers is to specify the horizontal type, which is designed to operate full of water.

Alternating steam and then water service must be avoided because chlorides in the water may be concentrated during the steam cycle.

Avoid SCC under thermal insulation. Another area where chlorides can concentrate is under thermal insulation. Whenever thermal insulation is required on hot austenitic stainless steel pipe, the probability of the insulation becoming wet should be carefully considered by the designer. Thermal insulation may contain water-soluble chlorides that can be leached out and deposited on the hot metal pipe. The consequent combination of high temperature, chlorides, and stress can result in SCC. Once, a large chemical company had an epidemic of such failures. One of these failures involved AISI 304 stainless steel pipe under wet 85% magnesia insulation, as shown in Figure 3.68. More failures persisted until a program of pretesting of various kinds of thermal insulation was completed.

FIGURE 3.68 — *AISI 304 stainless steel pipe that failed under thermal insulation by chloride stress corrosion cracking.*

In the pretest[37] (called *the wick test*), distilled water is wicked up from a dish to an electrically heated surface of a stainless steel U-bend specimen by a block of insulation material under examination, thus simulating the action of wet insulation on a hot stainless steel pipe. Figure 3.69 shows a U-bend specimen that failed this test by cracking. Many types of insulating materials have been subjected to this simulation of service conditions. Most types and brands of insulation were found to crack austenitic stainless steel in a short time; in several cases, insulation was found to crack austenitic stainless steels in as little as three weeks. Only two makes of insulation did not cause cracking. It was found that the addition of sodium silicate to the insulation mix of both insulations by manufacturers inhibited the cracking.

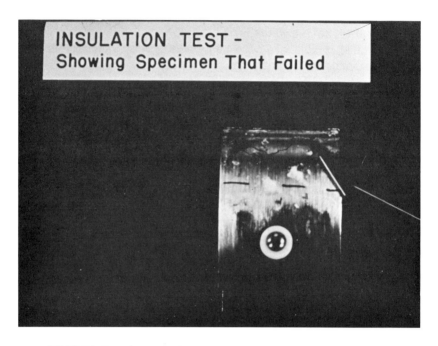

FIGURE 3.69 — *Specimen from a Dana and DeLong wick tester. Cracks pointed by arrow demonstrated that the insulation tested would crack austenitic stainless steel.*

It should be emphasized that the use of thermal insulation without chlorides will not necessarily eliminate SCC. The chlorides can be introduced from the water that soaks the insulation.

The designer is cautioned that this sort of cracking problem with insulated stainless steel pipe is not limited to just outdoor pipe. There are examples of cracked *indoor* pipes when adjacent equipment was washed down and the insulation on the stainless steel pipe was inadvertently soaked with water.

Gaskets, lubricants, insulators, adhesives, sealing compounds, asbestos compounds, pipe identification bands, and any other nonmetallic materials that will contact austenitic stainless steels under moist, elevated temperatures conditions should not contain water soluble chlorides. One lesson learned from testing nonmetallic materials for use in a large plant using mostly austenitic stainless steel equipment was that practically every material has unsuspected chlorides present. There was one instance of this when AISI 304 stainless steel piping cracked under foam glass insulation. It was determined that the adhesive used to attach the glass contained large amounts of water-soluble chlorides and that is where the cracking occurred. In another instance, polyvinylchloride (PVC) plastic identification bands were specified for miles of stainless steel piping. Cracking beneath these bands occurred when sunlight was present; this caused sufficient temperature to degrade the PVC and release chlorides. The cracking was quite superficial, affecting only a very shallow depth. In no instance were the walls penetrated, but there was no doubt that SCC was involved and probably severe cracking would have resulted after a long period of time.

The designer should be wary of all nonmetallic materials that are to be used in his designs for use in contacting austenitic stainless steel under heated and moist conditions. For critical areas, he should specify that there shall be no water soluble chlorides present in the required nonmetallics. Since suppliers often do not know the concentration of chlorides in their products, the designer should request that the suppliers prove that there are no soluble chlorides present by conducting appropriate tests.

The problem of SCC under insulation can also occur under unexpected conditions. For example, cracks were observed on an insulated and steam-traced AISI 309 stainless steel pipeline used to convey sulfur trioxide. These cracks appeared only at locations where there were pipe supports. The thermal insulation used was adequately inhibited by sodium silicate; however, the insulating cement used for making the joints at the hangers was loaded with chlorides without inhibitors, which caused the cracking.

Eliminate dead spaces. An additional way of avoiding SCC is for the designer to eliminate, as much as possible, in his stainless steel designs, crevices, dead spaces, pockets, blind flanges, etc., in which chlorides can concentrate.

Create compressive stresses. Another tool a designer can use to control SCC is shot peening. The surface of affected parts to be protected can be shot peened for greater resistance to SCC. Tensile stresses are required to cause SCC. Since shot peening places the surface of the treated metal in compression, SCC cannot occur. Compressive stresses can also be introduced on metal surfaces by other means, such as rolling threads on fasteners instead of machining them. Rolling gives added SCC resistance as compared to machine threads. Swaging also forms compressive stresses on the surface of parts. (Note: A system with compressive stresses will obviously also have tensile stresses. Such stresses may not be on the surface in contact with the corrodent, but the designer should assure that no damaging tensile stresses will inadvertently be left which will contact the corrodent.)

Use inhibitors. The use of inhibitors has been tested for controlling SCC in austenitic stainless steels. Among those tested were sodium metasilicate (mentioned previously as an inhibiting agent for thermal insulation), sodium nitrate, and an alkaline-phosphate water treatment. Other inhibitors mentioned in the literature include acetates, iodides, and benzoate ions. Some of these inhibitors may have been successful in delaying SCC, however, no inhibitor by itself should be relied upon to eliminate SCC.

Tests performed at the Du Pont Engineering Lab by D. Warren indicated that the addition of 1500 ppm of sodium metasilicate to water with a chloride content of 1500 ppm will materially delay cracking. The control specimens cracked in 72 hours, while the specimens in the inhibited water did not crack in approximately four weeks. The alkaline-phosphate water treatment has been used to delay cracking of austenitic stainless steel in a nuclear power steam system. This treatment plus oxygen scavenging with sodium sulfite have been used successfully to delay the initiation of cracks in several large heat exchangers. (See Chapter 6.5 for further information on inhibitors.)

Use cathodic protection. There has been limited use of cathodic protection for mitigating SCC. For instance, at a Canadian plant, lead was successfully used to protect AISI 304L stainless steel tubing from SCC. Cathodic protection should be considered only in critical applications when other means of protection are not practical. This protection method should not be used in service in oil refineries or other services prone to hydrogen induced cracking. (See Chapter 6.3 for a complete discussion of cathodic protection.)

Use protective coatings. Another form of protection is chemical-resistant coatings. One method used to prevent SCC of stainless steel under thermal insulation that may become wet is to paint the surface of the stainless steel with a protective coating. One paint system that has been used successfully for such an application was a high-build epoxy.

Silicone high-temperature paint has also been used. Any paint barrier that can prevent contact of the corrosive agent with the stainless steel will stop SCC, as long as the paint completely covers and is resistant to the environment. Protective coatings appropriate to the environment involved can be a low-cost remedy for SCC.

Reduce hardness. When possible, particularly on high-strength parts, the hardness should be held to the lowest possible value.

Avoid mercury. The designer should avoid specifying mercury-containing equipment, such as manometers, for determining pressure drop measurements, etc. When mercury-containing equipment must be used, the designer should specify a catch pot or another protective device to keep the mercury from inadvertently getting into process streams.

Carefully clean titanium. When specifying titanium for high-temperature service, make sure that parts are thoroughly cleaned of chlorides (even fingerprints) to avoid SCC. Use a solvent that contains no chlorides, of course.

☐ 3.9 Hydrogen Induced Cracking, Hydrogen Embrittlement, and Liquid Metal Cracking

Hydrogen Induced Cracking Mechanism

Hydrogen induced cracking (HIC) is a process similar to SCC that falls within the general category of environmentally caused cracking. Although the actual mechanism that causes HIC has not been established, it is known that the introduction of atomic hydrogen into a stressed metal triggers a condition which causes cracks to form. As opposed to normal corrosion where the degradation occurs at the anode; when HIC occurs, the degradation occurs at the cathode where hydrogen is generated.

As discussed in Chapter 1.4, during the corrosion process in an acid electrolyte, hydrogen is reduced from a valence of plus one (H^+) to atomic hydrogen with a valence of zero (H^0) at the cathode. Usually, the atomic hydrogen combines in pairs to form molecular hydrogen ($2H^0 \rightarrow H_2$). In this form, it can be harmlessly evolved as a gas. When certain materials called *poisons* are present at the cathode as the reduction process is proceeding, the uniting of pairs of hydrogen atoms is suppressed. The atomic hydrogen then is free to diffuse into the interstices (space between grains) of the metal. At this point, it is not definitely established just what causes atomic hydrogen to crack the metal; however, it is known that the metal must be stressed. Poisons include cyanides, selenium compounds, and hydrogen sulfide. The most aggressive poison appears to be hydrogen sulfide. When hydrogen sulfide is the poisoning agent, the cracking process is sometimes called *sulfide stress cracking* (SSC).

Where HIC occurs. HIC occurs in most high-strength materials like alloy steels, cold-worked austenitic stainless steels, ferritic stainless steels, and to a limited extent, even in some nickel alloys. These metals must be in a highly stressed condition. When a soft ductile material like low carbon steel is subjected to atomic hydrogen, the result is blistering or delamination. The petroleum industry is plagued with HIC of medium carbon steels because of the prevalence of what is termed *sour services*, which involves wet hydrogen sulfide (H_2S). Other industries experience this form of attack too when H_2S is present to act as a poison. These would include a waste water system in which waste organic products decompose to H_2S, gas scrubbing systems in which H_2S gas is present, and heavy water plants in which H_2S is present as a carrier. The most hazardous temperature for HIC is room temperature.

Hydrogen Embrittlement

Hydrogen embrittlement (HE) is related to HIC in that atomic hydrogen diffuses into the matrix of steel. When saturation occurs, only temporary embrittlement or loss of ductility of the steel occurs. It can be completely restored by heating at about 325 F (162 C) for about 1 to 3 hours.

Liquid Metal Cracking

Liquid metal cracking (LMC) is very similar to HIC in that certain metals with high tensile stresses are cracked by specific species, in this case liquid metals. This can be an aggravating problem, especially during welding. Later in Chapter 4.1, an example is given in which AISI 316 stainless steel sheet has become embrittled by being welded to a galvanized angle. In this case, molten zinc was the agent that cracked the stainless steel. There are other combinations of molten metals that can crack certain metals, as summarized in Table 3.7.

As also pointed out in Chapter 4.1, when welding materials like galvanized steel or copper-clad steel, all materials that can trigger LMC must be removed by acid, machining, or by other means. Grinding is not usually a good removal method because it smears ductile materials which then will not be entirely removed.

TABLE 3.7
Liquid Metal Cracking Combinations[1]

Stressed Metal	Molten-Cracking Agent
Yellow Brass	Mercurous Nitrate (metallic mercury)
Copper Alloys	Mercury
Alloy 400	Mercury
High-Strength Aluminum Alloys	Sodium Tin Zinc
Steel	Copper (welding of copper-clad steel) Tin Cadmium Zinc Lead
Austenitic Stainless Steel	Zinc Aluminum Lead
Titanium	Cadmium

[1]From Reference 33.

As mentioned before, mercury from blown manometers or from some catalysts must be controlled when stressed metals are involved or LMC may result. When cadmium-plated parts or parts covered with aluminum or lead are to be welded, the covering materials must first be removed from the joint area before welding.

What the Designer Can Do to Prevent HIC, HE, and LMC

The following are steps and considerations a designer can take or act upon to prevent HIC, HE, and LMC:

Specify anodic protection. When HIC is likely, the designer can specify *anodic protection* (Chapter 6.4) to alleviate cracking. However, never specify cathodic protection because this kind of protection system increases activity at the cathode creating more hydrogen, hence accelerating cracking.

Be cautious about ASTM A-193 Grade B-7 bolting. When specifying high-strength bolting for such items as pressure vessels and pipelines where 90,000 to 100,000 psi minimum yield strength is required (such as with spiral wound gaskets), be cautious about specifying ASTM A-193 Grade B-7 bolting. This specification has become a standard practice over the years for high-strength service. This grade of bolt is made from AISI 4140 steel hardened and tempered at 1100 F (593 C) for a Rockwell C Hardness (HRC) of about 35. It has been determined from many test programs that HIC will not occur in environments if the HRC reading of the metal is below 22 and the bolts are not loaded past their yield strength.[16] The designer should specify the controlled hardness grade of ASTM Grade B7M, which has a maximum hardness of 22. It was developed for sour crude service. Strain-hardened AISI 316 stainless steel studs to ASTM A-193 B6M Class 2 will also resist HIC, but these are more expensive.

Specify more resistant materials. When possible, specify the more resistant materials, such as Alloy 20 or Hastelloy C-276.

Specify heating after electroplating or pickling steel. When specifying that steel is to be electroplated or pickled, the designer should also specify that the parts be heated at 325 F (162 C) afterwards for one to three hours to expel the hydrogen that has saturated the metal after the operation.

Specify the removal of coatings on steel before welding. Specify that before welding, all galvanizing (zinc), cadmium, copper, or other coatings on steel be removed by acid or machining (not grinding).

☐ 3.10 Corrosion Fatigue

Corrosion Fatigue Mechanism

Corrosion fatigue may be defined as a combination of normal fatigue and corrosion that causes failure at a tensile stress way below the normal endurance limit of the metal involved. When a part fractures by straight fatigue, a repeated or cyclic stress must have been applied that is above the endurance limit of the metal. Figure 3.70 shows a stress vs number of cycles graphs called *SN curves*. Curve A represents an SN graph of a metal in air. Note that no matter how many cycles are applied, if the stress is below the endurance limit, no fracture will occur. However, if the same metal is exposed to a corrodent, such as fresh water or salt water and the same stress, the endurance limit will be drastically lowered, as represented by Curve B and failure (cracking) can occur. When corrosion fatigue is in progress, it will produce a failure

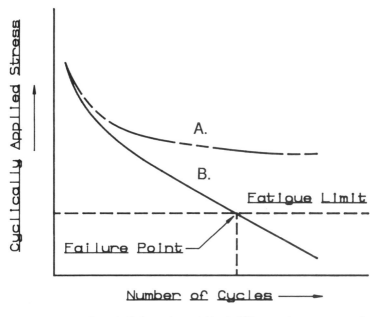

FIGURE 3.70 — A. Fatigue of a metal in air S/N curve: (— — — —)
B. (————) same metal exposed to a corrodent.

after some finite number of cycles regardless of how low the applied tensile stress may be because of the continuously descending SN curve.

Compressive stresses will not cause corrosion fatigue.

Corrosion fatigue is similar to SCC, except that the stress involved is cyclic or repeated. This type of corrosion is a serious cause of failure when such stresses are created, as in reciprocating machinery under corrosive conditions. The petroleum industry, in particular, has suffered from corrosion fatigue. For instance, drill pipe and sucker rods submerged in brines and sour crudes fail often because of this corrosion mechanism.

Examples of Corrosion Fatigue Failures

Below are field examples of corrosion fatigue failures.

> *Example 1:* The materials engineering for a vertical pump designed for submerged water service had been well done, except for one component. The shaft was stainless steel,

and the volute and check valve assembly were made of polyvinyl chloride, but the vertical column that housed the shaft and supported the motor was made of low-leaded brass. A torque was applied to the column every time the motor started. This was repeated many times during the day. The column failed by corrosion fatigue in a characteristic torque pattern, as shown in Figure 3.71. The pump was replaced by a similar one with an austenitic stainless steel column, which has lasted five years and still has not failed.

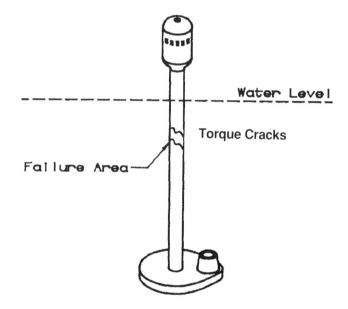

FIGURE 3.71 — *Submerged pump; vertical column failed by corrosion fatigue.*

Example 2: A well-documented corrosion fatigue failure case[38] occurred during World War I when steel tow lines for mine sweeping operations repeatedly fractured. The seawater was the corrosive media and the vibration of the steel lines under tension as they pulled the paravanes through the water created the cyclic stress. Substituting higher strength steel wire did not solve the problem; however, the use of cathodic protection in the form of galvanized steel wire did stop the failures. (See Chapter 6.3 for further information on cathodic protection.)

Example 3: More recently in 1981,[39] the first stage blade of an axial air compressor in an acid plant failed. The blade was made of AISI 403 stainless steel covered with a protective coating. This consisted of an inorganic coating formulation with aluminum added as a filler. The air handled contained ammonium sulfate and phosphate, and nitric acid fumes along with moisture. This atmosphere was known to pit AISI 403 stainless steel aggressively. An investigation disclosed that corrosion fatigue had started in areas where the coating had deteriorated. Corrosion pits that had formed on the AISI 403 blade had triggered the corrosive action. Apparently, the coating had failed after about six years of service. The solution to the problem by the plant involved was to have the blades recoated by the vendor periodically.

What the Designer Can Do to Reduce Corrosion Fatigue

The following are guidelines for a designer who may be dealing with a corrosion fatigue situation.

1. Specify a more resistant material.
2. Change the the design to dampen or stop the stress cycling.
3. Eliminate the corrosive condition by relocating the equipment, when possible.
4. Avoid stress raisers by avoiding abrupt changes in cross section. Provide generous fillets and avoid sharp corners. Use tapered sections instead of stepped changes. Avoid holes and attachments when possible. If this is impossible, place them in low stress areas.
5. Use butt welds instead of fillet welds. Fillet welds are much more susceptible to corrosion fatigue than butt welds.
6. Avoid rough machining and grinding, as well as sheared and punched edges, since they can initiate corrosion fatigue. The edges should be machined and polished.
7. Shot peen the part. This puts the outer surface in compression and since fatigue requires tension stresses, corrosion fatigue should not occur.

8. Use inhibitors. If the part is submerged in a corrodent such as a reciprocating arm operating in a brine, inhibitors have proved effective in preventing corrosion fatigue.
9. Use galvanic protection.
10. Improved corrosion fatigue properties will also result if the part involved is nitrided or carburized.
11. Specify that there be no decarburization if the part is to be heat treated, since this increases susceptibility to corrosion fatigue.
12. Specify that an appropriate protective coating be applied to the part. The designer's specification should instruct that the part must be inspected regularly and recoated when necessary.

☐ 3.11 Intergranular Corrosion

Intergranular Corrosion Mechanism

When metals are cast, the solidification begins at randomly distributed areas. Growth from these nuclei continues in a regular atomic array to form so-called "grains." Because of the randomness of solidification, when one emerging grain meets another, an irregular grain boundary is formed. These grain boundaries are sometimes preferentially corroded. When this occurs, it is called *intergranular corrosion* (IGC).

IGC is caused by impurities or alloying elements that have migrated from the surrounding areas to the grain boundaries and then precipitated between the grains. These precipitated materials have a different corrosion potential than the adjacent grains. They can be either cathodic or anodic to the surrounding area, depending on the nature of the precipitate and the metal involved.

Certain environments will cause IGC in certain metals. When these metals are in these environments, if the precipitate is anodic, the precipitate will corrode; if the precipitate is cathodic, a narrow area *next* to the grain boundary will corrode. However, regardless of whether the precipitate is anodic or cathodic, the result is about the same. A channel or crack will form around the grains and have a deleterious influence on the mechanical properties of the metal.

Another way to cause precipitates to form between the grains of a metal is by severe cold working. Rolling plates and sheets in a mill can cause an elongation of the grains parallel to the surface in the direction of rolling. In such cases, IGC can occur in hostile environments and when the corrosion products expand, *layer corrosion* or *exfoliation* can occur. In severe cases, the edges of the metal resemble separated pages of a book. Aluminum alloys are most susceptible to exfoliation.

Along with aluminum alloys, IGC occurs most often in copper alloys and 300 Series stainless alloys.

Examples of Intergranular Corrosion

The following are examples of IGC:

1. Austenitic stainless steels in some media are susceptible to IGC, which is related to the formation of chromium carbides in the grain boundaries. These carbides are formed when the stainless steel is heated in the sensitizing range of 900 to 1400 F (482 to 760 C). One explanation of IGC is called *the Chromium Depletion Theory*. This theory states that when the chromium carbides are precipitated, chromium lean areas are formed next to the grain boundaries, thus rendering these areas less corrosion resistant. Figure 3.72 shows a striking example of this

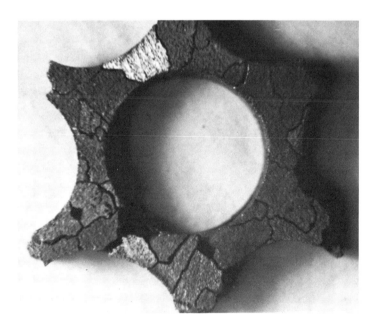

FIGURE 3.72 — *Part of badly attacked tube sheet made of cast 18Cr-8Ni. The IGC was caused by improper heat treatment.*

form of attack. A cast 18Cr-8Ni tube sheet was used in hot nitric acid service. The tube sheet had been stress relieved at a temperature where carbon steel is normally stress relieved at 1150 F (620 C) for one hour per inch (25 mm) of thickness, with a minimum of one hour. Such a heat treatment in the sensitizing range assured completely that chromium carbides would be precipitated. The result was that the tube sheet was so badly attacked that it literally fell apart.

2. Figure 3.73 shows a photomicrograph at 50X of a piece of AISI 316 stainless steel that had experienced IGC. Note the intergranular nature of the failure. At the bottom of the field, as shown in Figure 3.73, entire grains became loose and fell out.

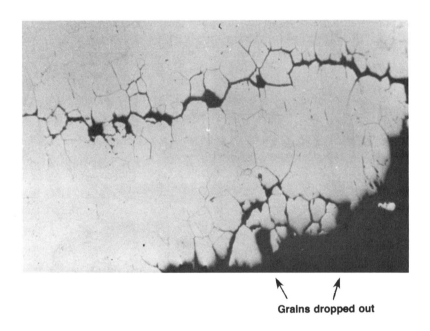

Grains dropped out

FIGURE 3.73 — *Intergranular attack in AISI 316 stainless steel. (Original magnification was 50X.)*

3. Figure 3.74 shows a sectioned AISI 304 stainless steel elbow that had become sensitized and failed in 30% nitric acid at 200 F (93 C) by IGC. Note that the treaded sections were not affected because they were covered up by the AISI 304L stainless steel piping and did not contact the acid.

FIGURE 3.74 — *Sensitized AISI 304 stainless steel elbow that failed by IGC; note that the threaded sections were unaffected.*

4. Figure 3.75 shows an AISI 304 stainless steel welded tee that corroded adjacent to and in back of welds. In this case, the areas next to the welds were sensitized by the heat of welding, and chromium carbides were precipitated causing excessive corrosion in nitric acid service.
5. Figure 3.76 shows an AISI 309 stainless steel Schedule 40 pipe welded to a welding flange. The weld areas and heat affected zones were sensitized and were consequently corroded by IGC in nitric acid service. Note the loss of metal adjacent to these areas. As a matter of fact, many years ago the ASME Boiler Code required a stress-relief in the sensitizing range for austenitic stainless steels. (See Chapter 5.1 for descriptions of ASME and other organizations.) This stress-relief was the heat treatment designed specifically for carbon steel. Many sensitized pieces of stainless steel process equipment consequently failed by IGC when used in service that could cause this kind of corrosion after being subjected to that heat treatment. This problem was corrected by the ASME Boiler Code years ago by specifying an appropriate heat treatment. (see Chapter 4.2.)

FIGURE 3.75 — *AISI 304 tee. IGC occurred behind the weld areas.*

FIGURE 3.76 — *AISI 309 Schedule 40 flanged pipe. IGC also occurred behind the weld areas here.*

Note: The heat treatment that will place the precipitated chromium chlorides back into solution in sensitized unstabilized grades of austenitic stainless or grades of austenitic stainless steel with carbon contents above 0.03%, is to hold at 1900 to 2000 F (1038 to 1093 C) and quench to black heat within 3 minutes. This is appropriately called a *solution heat treatment* and will make the stainless steel immune to IGC. (See Chapter 4.2.)

    6. Other metals and alloys are also subject to IGC. For instance, Figure 3.77 shows a photomicrograph at 100X magnification of nickel that had experienced a high heat in the presence of sulfur causing IGC. Note the pronounced corrosion between grains and loss of whole grains in the center of the field.

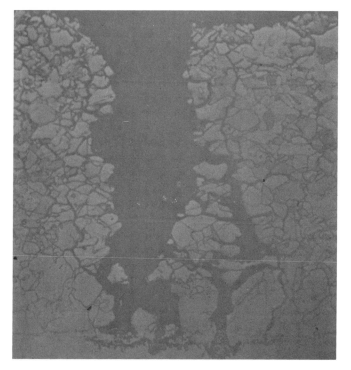

FIGURE 3.77 — *Nickel part that failed because of embrittlement in the presence of sulfur at high temperature (100X magnification).*

Effects of Stabilizers and Low Carbon Content on Intergranular Corrosion of Stainless Steels

When stainless steels were first available commercially in the 1920s, they were considered to be the panacea for all corrosion problems, but then they started to fail by IGC. The search for a way to eliminate IGC is an interesting story. The problem was solved by adding either titanium or niobium (columbium) to the stainless steel melt. The carbon in the steel has a greater affinity for titanium or niobium than it does for chromium. Therefore, either titanium carbides or niobium carbides are formed and precipitate harmlessly in the grain boundaries in place of the chromium, which then stays in solid solution, which consequently preserves the corrosion resistance. These so-called "stabilized" steels were called AISI 321 (stabilized with titanium) and AISI 347 (stabilized with niobium). (See Chapter 5.2 for further details.)

For years, AISI 321 and 347 stainless steels were used extensively in most industries to avoid IGC. Later, a low carbon stainless steel was developed, which accomplished the same job as the stabilized grades. It was found that if the carbon content was limited to 0.030% carbon, there would be insufficient carbon to form with the chromium to cause chromium lean areas to be formed. These steels are called the *extra low carbon grades* and are made up of steels such as AISI 304L and 316L steels. Extra-low carbon grades have substantially replaced the stabilized grades in the chemical industry, except for use in high temperature service.

Knife Line Attack

At one time, a phenomenon called *knife line attack* was not clearly understood and was considered a separate class of corrosion. Deep grooves would appear adjacent to both sides of a weld (as though a knife had made a cut) without affecting the weld metal or other areas of the weldment. Figure 3.78 shows an AISI 304 stainless steel specimen cross-welded with an AISI 308 stainless steel shielded-arc electrode. Figure 3.79 shows a similar specimen, only the material was the low carbon stainless steel grade AISI 304L steel welded with an AISI 308L steel electrode. Both specimens were exposed to 65% boiling nitric acid for 48 hours. As can be seen, the AISI 304L steel specimen was essentially untouched, while the other one showed knife line attack, which is actually intergranular corrosion. The explanation is that during welding, the areas adjacent to the cross welds of the AISI 304 stainless steel specimen were held in the sensitizing range long enough to precipitate chromium carbides. The reason the weld did not corrode was that right after the weld had been completed, the temperature was above the sensitizing range and when it cooled, it cooled so rapidly past the carbide precipitation range that the carbides did not have time to precipitate. The parent metal away from the weld did not corrode because this area did not reach the sensitizing temperature range. The AISI 304L steel specimen was not attacked because its carbon content was below 0.030%.

FIGURE 3.78 — *AISI 304 stainless steel cross welded with an AISI 308 shielded-arc electrode and then exposed to 65% boiling nitric acid for 48 hours. (Note the intergranular corrosion.)*

Environments That Can Cause Intergranular Corrosion

Many acids can cause IGC in austenitic stainless steels. Table 3.8 shows the results of tests[40] conducted on eight of these acids. As can be seen, the acid concentration and temperature must be of specific values for this type of corrosion to occur.

Allessandria[41] reporting on case histories of refinery experience with the austenitic stainless steels outlines a case where rapid

FIGURE 3.79 — *AISI 304L (0.03% carbon content) cross welded with an AISI 308L electrode and also exposed to 65% nitric acid for 48 hours. (Note that there is no intergranular corrosion present.)*

intergranular failure resulted from polythionic acids ($H_2S_xO_6$) contacting sensitized AISI 309 stainless steel. The polythionic acids were formed during shutdown periods by the reaction between iron sulfide scale and moisture.

Many other solutions will cause IGC in austenitic stainless steels under specific conditions, including:

>Ammonium Nitrate
>Calcium Nitrate
>Chromium Chloride
>Copper Sulfate
>Ferric Chloride
>Ferric Sulfate
>Sodium Hypochlorite

TABLE 3.8
Intergranular Corrosion of Stainless Steel[1]

Solution	Concentration (%)	Room Temperature	140 F	Boiling
Nitric Acid	1	0[2]	0	X[3]
	10	0	0	X
	30	0	0	X
	65	0	X	X
	90	X	X	X
Lactic Acid	85	0		X
Sulfuric Acid	0.5	0		X
	2	0		X
	10		X (158 F)	
	30	X		
	95	X		
Acetic Acid	30	0		
	99.5 (glacial)	0		X
Formic Acid	10	0		
	90	0		X
Chromic Acid	10			X
Lactic Acid	50			X
Phosphoric Acid	60			X
	70			X
	85			X

[1] Exposure time was 40 weeks; specimens were sensitized AISI 304 and 316 stainless steels. (From Reference 40.)
[2] 0 = test conducted but no intergranular corrosion.
[3] X = intergranular attack.

The designer should remember that many environments will not affect austenitic stainless steels, even if they are sensitized and chromium carbides have precipitated. However, he should be aware of those applications in which IGC may occur.

Testing for Intergranular Corrosion Susceptibility

Several tests have been devised to evaluate the susceptibility of individual heats of austenitic stainless steel to IGC. One of them, termed *the boiling nitric acid test*, has been useful over the years in establishing that solution-annealed heats of unstabilized steels that have passed the test have not been sensitized and that stabilized and extra low carbon grades will not become sensitized when held for one hour at 1250 F (677 C). (See ASTM A262 Practice C, "Detecting Susceptibility of Intergranular Attack in Stainless Steels," for details of this test.)

The ASTM A262 Practice C test is not intended to measure the corrosion resistance in any other media other than the conditions of the test. A low nitric acid rate, however, means that the stainless steel would not be expected to be susceptible to IGC in other solutions.

What the Designer Can Do to Eliminate Intergranular Corrosion

The designer can take the following steps to eliminate IGC:

Consider the metal and the process combined. Ascertain that the metal selected, whether an aluminum alloy, a copper alloy, or a stainless steel, etc., is resistant to IGC in the process being considered.

Consider the metal's microstructure. When considering using metals in IGC-causing environments, assure that the microstructure of the metal is immune to IGC. A proper heat treatment or limiting cold work may accomplish this.

Do not specify unstabilized grades of stainless steel. When process conditions can or are suspected to cause IGC in stainless steels, specify either extra-low carbon steel, such as AISI 304L, or the stabilized grades, such as AISI 321 or 347 steel, rather than any unstabilized grades. Usually, the cost of specifying an unstabilized stainless steel (AISI 304, 316, 309 steels, etc.) and heat treating the finished weldment is more expensive than using a stabilized or low carbon type (AISI 347, 321, 304L, 316L, 309L steels) that does not require heat treatment.

Specify the 65% boiling nitric acid test. When a critically corrosive condition where an austenitic stainless steel has been found to be resistant is involved, specify that specimens from all heats of steel offered for such use undergo the 65% boiling nitric acid test. (See ASTM A262 Practice C for details.) This will assure that such heats have been produced and heat treated so that their maximum corrosion resistance has not been impaired.

Specify proper welding electrodes or filler metal. Make sure that the proper welding electrode or filler metal (either extra low carbon or stabilized) is specified.

Specify a solution heat treatment. When heat treating unstabilized stainless is required, specify a *solution heat treatment*. (See Chapter 4.2.)

Avoid stress relieving austenitic stainless steels at carbide precipitation range temperatures. Do not allow stress relief of austenitic stainless steels at temperatures in the carbide precipitation range [900 to 1400 F (482 to 760 C)].

Specify the stabilizing heat treatment for austenitic stainless steel stress relieving. Generally, austenitic stainless steels do not require relief of mechanical stress to maintain structural stability because of their inherent ductility. When such a heat treatment is demanded, however, use the stabilizing heat treatment. (See Chapter 4.2.)

3.12 Dealloying

Dealloying Mechanism

Dealloying causes the selective removal of one constituent from an alloy while leaving the original dimensions essentially unchanged. However, the affected areas are left very weak. Dealloying is also called *parting corrosion* and is occasionally called *metasomatic corrosion*, *selective leaching*, or *demetallification*. For example, in certain environments such as in contaminated water, brass will corrode in this fashion. This is termed *dezincification*. In essence, the zinc is completely removed from an area, leaving only copper. Actually, it is thought that all brass containing zinc and copper is dissolved, but the copper is redeposited by electrochemical deposition in the cavity that had just been formed by the corrosion. The color of the deposited mass is reddish or copper-colored in contrast to the yellow, unaffected alloy. An example of *plug-type dezincification* is shown in Figure 3.80. The susceptibility of brasses to this form of corrosion is increased by high chloride and high carbon dioxide contents, high temperatures and low velocity of the water plus crevices and inert deposits on the metal. Usually, dezincification does not occur in brasses containing less than 15% zinc. This dealloying may be controlled by adding tin, phosphorus, or antimony to brass containing 15% or more of zinc. Some copper zinc alloys, such as admiralty metal, are inhibited by such additions. For instance, inhibited admiralty metal generally contains 70% copper, 29% zinc, 10% tin, and small amounts (0.02 to 0.06%) of phosphorus, arsenic, or antimony.

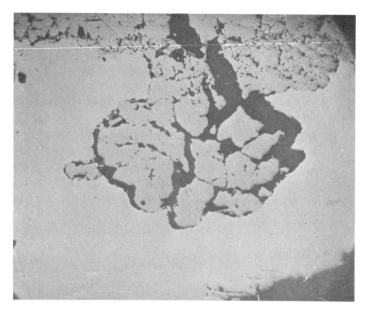

FIGURE 3.80 — *Plug-type dezincification. (Original magnification was 25X.)*

Examples of Dealloying

Under certain hot environments containing hydrogen chloride, chromium can be selectively removed from Alloy 600 This is called *dechromification*. At one plant where Alloy 600 tubes are used for such an environment, magnets are effectively used to determine what areas have lost chromium.

Copper-aluminum alloys can experience a loss of aluminum; this is called *dealuminification*.

Cobalt may be removed from Stellite #1 (CoCrWC alloy) by sulfuric acid slurry containing ferrous sulfate. This is called *decobaltification*.

Graphitization of cast iron is a form of corrosion where the iron is simply corroded out, leaving the graphite still in place. This form of attack occurs in soils containing sulfates or sulfate-reducing bacteria. Graphitization of cast iron is also prevalent in acid waters and salt water.

Other examples are the removal of tin from lead-tin solders, silicon from silicon-copper alloys, copper from copper-silver alloys, nickel from copper-nickel alloys, and silver from silver-gold alloys.

What the Designer Can Do to Reduce Dealloying

Below are actions a designer can take to reduce dealloying.

1. Carefully consider the environment involved and select the proper materials. Probably the greatest offender for causing dealloying is water. The designer should determine in advance how aggressive the water will be by having appropriate tests conducted.

2. Brasses probably fail the most by this type of corrosion. If there is a history of such failures or tests that disclose a weakness, other materials should be considered. For instance, in brackish waters, in ascending order of resistance to dealloying, bronzes would be considered first followed by aluminum bronzes, copper-nickel alloys, and Monel. Under certain circumstances, which will be discussed in Chapter 5.2, austenitic stainless steels can be considered for use in brackish waters.

3. When selecting copper-base alloys for water service, the designer should select those with less than 15% zinc to avoid dealloying.

4. When brasses are required for water service for economic reasons, avoid:

    —high water temperatures

    —high chloride content of the water (above 500 ppm)

    —crevices

    —low water velocity [below 1 ft/s (18 m/min)]

    —high carbon dioxide content of water

    —deposits on surface of brass

    Note: Use another material of construction if any of the above cannot be avoided.

5. When cast iron is to be specified for burial in the soil, specify a cast iron with at least 2 to 3% nickel to avoid possible graphitization.

3.13 Hydrogen Grooving

Hydrogen Grooving Mechanism

Hydrogen grooving (HG) is a type of corrosion peculiar to the storage and handling of strong sulfuric acid in carbon steel. This can occur in storage tanks, pipelines, and tank cars.

When carbon steel corrodes, ferrous sulfate is precipitated on the steel surface as a soft film or layer and hydrogen is evolved as a gas:

$$Fe + H_2SO_4 \rightarrow FeSO_4 + H_2 \qquad (4.1)$$

The ferrous sulfate layer acts as a corrosion barrier and significantly reduces the corrosion rate. The published corrosion rates are low. Figure 3.81 shows rates of 5 to 20 mpy at ambient temperatures between 65 and 96% sulfuric acid, while above 96%, the rates are between 0 to 5 mpy. Under special circumstances (when strong acid is not moving), the hydrogen evolved in the corrosion of steel in sulfuric

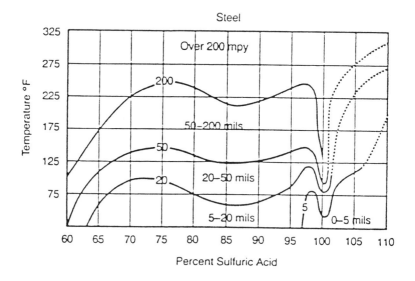

FIGURE 3.81 — Corrosion of steel by sulfuric acid as a function of concentration and temperature (reproduced with permission from M. G. Fontana, N. D. Greene, Corrosion Engineering, copyright 1978, McGraw-Hill, Inc.)

acid can cause considerable corrosion problems of high local attack termed *hydrogen grooving*. Bubbles of the gas are formed during the corrosion of steel as the iron is oxidized and the hydrogen is reduced. These bubbles ascend in the acid by a steady passage over the steel (covered with ferrous sulfate), which can disturb the ferrous sulfate and result in groove formation. This grooving can continue reducing the metal thickness at a rate much greater than the general corrosion rate of steel in the same acid concentration and temperature. The rate-determining factor in the corrosion of steel in strong sulfuric acid is the diffusion of ferrous ions into the bulk solution from the saturated solution at the film-acid interface.[42] If a sufficient quantity of hydrogen bubbles exists on a preferred path, the normal steady state of low corrosion will be disrupted and grooving will result with high local corrosion rates of 125 mpy or more.

Examples of Hydrogen Grooving

For horizontal tanks and pipelines (acid quiescent), the bubbles formed on the lower half of the circumference stream upwards until the upper half of the circumference is reached. At this area, the bubbles rise but are constrained to follow the contour of the upper half of the horizontal tank or pipe section. The corrosion rate in the grooves is greatly stimulated by the motion of the bubbles as they thin the diffusion layer and the protective ferrous sulfate protective film, as mentioned previously. After a path has been established, a steady stream of hydrogen bubbles follows the same path gradually deepening it. Figure 3.82 shows a cross section of the top 180 degrees in a 3-in.- (76-mm)-diameter pipe that had contained stagnant 60-degree Baume sulfuric acid. Note the deep groove in the top center. Figure 3.83[43] shows the top 180 degrees of a 3-in.- (76-mm)-diameter steel elbow that had been sectioned to show the HG. The grooves began at the 9 and 3 o'clock positions and traveled up the contour until the bubble streams converged at the top or 12 o'clock position. There, a much deeper groove was formed because of the increased hydrogen bubble density. Figure 3.84[43] shows a drawing of a top section of a steel pipe where a top center groove was made by hydrogen during nonpumping hours. Many horizontal nozzle and manways have failed by this top grooving action.

HG can also occur in vertical tanks. Figure 3.85[43] shows HG in the side wall of a vertical sulfuric acid tank. This corrosion resulted from the location of the acid inlet in the roof which was positioned too close to the side wall. HG combined with erosion-corrosion caused this deterioriza-

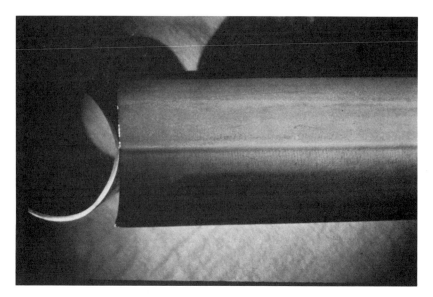

FIGURE 3.82 — Cross section of the top 180 degrees in a 3-in- (76-mm)-diameter steel pipe that had contained 60-degree Baume sulfuric acid during an extended plant shutdown.

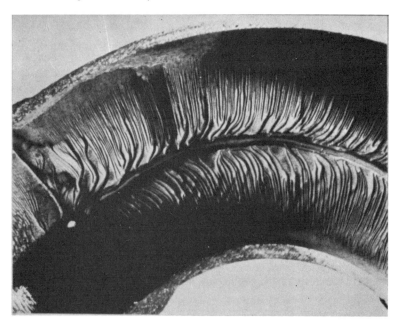

FIGURE 3.83 — Section of the top 180 degrees of a 3-in.- (76-mm)- diameter steel elbow showing hydrogen grooving caused by contact with static sulfuric acid.

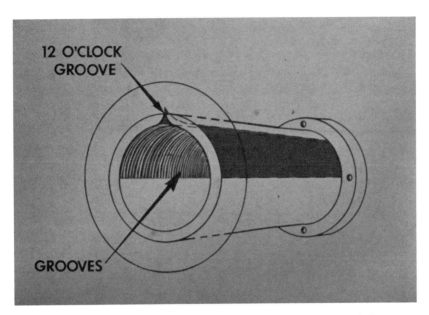

FIGURE 3.84 — *Top section of a steel pipe grooved by hydrogen during non-pumping hours of sulfuric acid.*

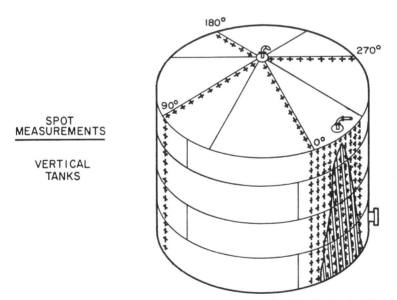

FIGURE 3.85 — *Drawing of a vertical sulfuric acid tank indicating location of hydrogen grooving on the interior side wall caused by acid inlet too close to the side wall.*

tion. Figure 3.86 shows an interior view of the side wall of a sulfuric acid tank just above the bottom of the tank. Note the severe grooving caused by hydrogen. Several cases of ruptured side walls of large tanks have occurred because of this type of corrosion.[44]

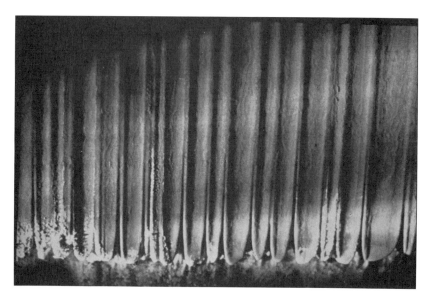

FIGURE 3.86 — *Hydrogen grooving of interior side wall just above the bottom of the sulfuric acid tank.*

What the Designer Can Do to Reduce Hydrogen Grooving

Vertical tanks. Guidelines for reducing HG in vertical tanks are as follows:

1. The top 180 degrees of manways and horizontal nozzles are susceptible to HG. See Figure 3.87 for these critical locations. Therefore, specify that the vulnerable top of the neck of manways be projected about 2 in. (51 mm) inside the interior of the tank. This visor or hood should be lined with plug-welded strips of aHG-resistant alloy. See Figure 3.88 which depicts good and poor designs of manways. Carpenter Alloy 20-Cb-3 or Hastelloy C-276 should be used for the strip material (or welding

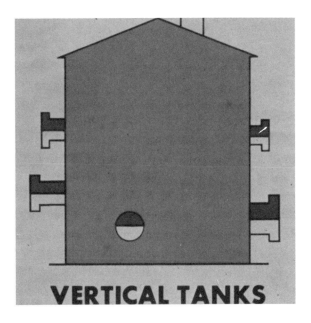

FIGURE 3.87 — *Critical locations of nozzles and manways in a sulfuric acid tank that should be protected from hydrogen grooving.*

overlays) for protection in 70 to 100.5% sulfuric acid. AISI 316 stainless steel may be used instead of the high-nickel alloys for concentration of 90 to 100.5%. No lining is required for acid strength above 100.5%

2. Locate roof acid inlet as far from the tank shell as possible, but at least 8 ft (2 m) away when tank size permits.
3. Include an outlet in tank designs to allow acid drainage before the acid contacts the roof. When overfilling occurs, HG can occur. When HG penetrates the roof, rainwater can enter the tank and dilute the water, and aggressive general corrosion can result. Weaker sulfuric acid with less specific gravity than strong acid will float on the surface and cause localized corrosion around the inside circumference of the tank.
4. Specify weld neck flanges instead of plate flanges when available to avoid an internal weld between the plate flange and the manway neck. Such welds in the top 180 degrees of a manway are very susceptible to aggressive HG attack.

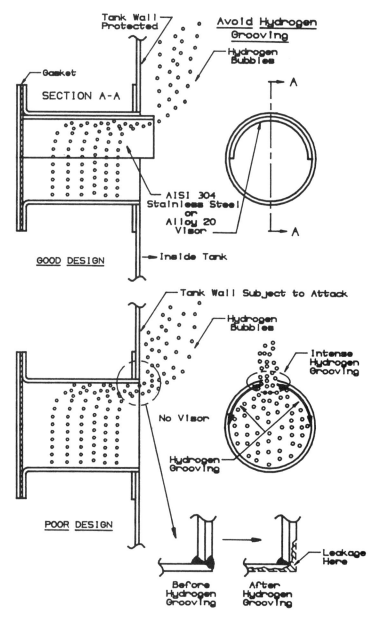

FIGURE 3.88 — *Good and poor design of manways in sulfuric acid tanks.*

5. Manway covers and blind flanges should be protected with a corrosion-resistant gasket such as Teflon.

Horizontal tanks and piping. Guidelines for reducing HG in horizontal tanks and piping include:

1. The top 180 degrees in horizontal tanks and rail cars up to the highest acid level are also susceptible to HG. See Figure 3.89 for critical areas. Specify that a protective internal lining be applied to solve this corrosion problem. Heat-cured phenolic linings have been widely used for 93% sulfuric acid and give a reasonable life in 98% acid. Hypalon[†], and of course glass linings, can also be used. (See Chapter 6.6 for protective coatings.)

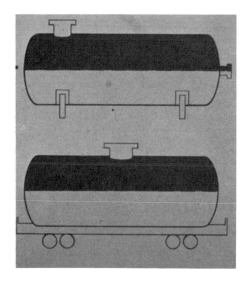

FIGURE 3.89 — *Areas susceptible to hydrogen grooving in horizontal tanks (top) and rail cars (bottom).*

2. HG occurs only when the flow in pipelines is practically zero and solar heating conditions exist. Therefore, long shutdowns should be as few as possible. This specification should be recorded in

---
[†]Registered trade name.

the designer's overall materials guide. (See Chapter 5.9.) Maintaining the velocity above 0.3 m/s (1 ft/s) in pipelines containing sulfuric acid during long shutdowns will avoid this problem. If this is impractical, specify lined carbon steel pipe such as polyvinylidene fluoride (Kynar[†]) or increase the corrosion allowance by specifying heavier walled pipe.
3. Sulfuric acid tanks should generally incorporate butt welds in side wall construction. However, where double-lap welds must be used, position them as shown in the "Better Design" in Figure 3.90.

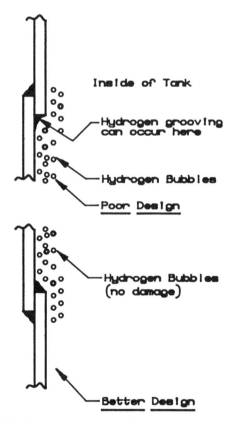

FIGURE 3.90 — *In side wall construction of sulfuric acid tanks where double-lap joints are specified, position lap as shown in "Better Design."*

[†]Registered trade name.

4 — DESIGN SOLUTIONS TO CORROSION PROBLEMS BASED ON FABRICATION TECHNIQUES AND ENVIRONMENTAL FACTORS

4.1 Welding

The Nature of Welding

Welding is used almost universally in the fabrication of process equipment. Over 90% of all permanent closures are made by fusion welding or brazing. For all its utility, welding has inherent characteristics that can foster corrosion. For instance:

1. The cast structure of a weld can be quite different from the usual wrought structure of the parent materials.
2. Weld spatter can create obstructions that can result in localized corrosion.
3. Many weld joints can contain crevices if not welded properly.
4. The weld surface is generally rougher than the parent material's surface.
5. Metal arc and submerged arc welding processes generate slag, which can set up corrosion cells.
6. Welding entails intense localized heat, which creates heat affected zones in the parent metal where phase transformations and precipitation can occur.
7. Welds contain internal shrinkage stresses.
8. Residue not removed from welding and brazing fluxes can be corrosive. Figure 4.1 exhibits an aluminum elbow that failed because the welding flux had not been removed after welding.

Although welding has some drawbacks, as noted above, it is still the best and the soundest method of closure available. Rivets, for instance, have built-in crevices and continuous joints are difficult to maintain. These problems are also experienced with bolting. In addition, the parent metal is seldom matched by the rivets or bolts.

FIGURE 4.1 — *Aluminum elbow that failed because the welding flux was not removed.*

Welding Decisions

Because of the problems of welding, the designer must assure that the welds in his structures are properly designed and specified. A mistake a designer can make is to simply note on his drawing that the structure "is to be welded." Unfortunately, this is standard procedure with some designers. Such a specification leaves the welding decisions up to the welder or the welding foreman who probably does not know what process conditions are involved. The end-use of the equipment must be carefully considered in advance and the appropriate weld design must be specified while the equipment is still in the design stage, not when it is already in the weld shop.

Welding Symbols

The type of weld joint expected should be noted on the drawing by the designer. According to this author's experience, not enough designers use American Welding Society (AWS) welding symbols. These symbols take the guesswork out of interpreting what joint is required, thus hopefully assuring optimum corrosion resistance. The symbols provide an easy and simple way of providing complete welding information on drawings. Specialized equipment is often designed by a contractor rather than by a company designer. The designer who arranges for such equipment to be fabricated should carefully check the contractors' drawings before giving his approval for fabrication. Another reason why a designer should know AWS symbols is because most competent equipment builders use these symbols and the designer should of course understand them.

The key to AWS welding symbols is the bent arrow that points to a weld joint on a drawing of a weldment. Each weld joint has its own arrow. The arrow has a head, a shaft called a *reference line*, and a tail. A number of symbols denote certain types of welds. If this symbol is placed below the reference line, the type of weld applies to the side of the joint nearest the arrowhead. If the arrowhead is placed above the reference line, the type of weld symbolized is to be located on the other side of the joint. Figure 4.2[45] displays the basic welding symbols and their location's significance. Five groove joints are shown, the square groove (which requires no preparation), the V-bevel groove, and the U- and J-grooves. A fillet, plug or slot weld, and weld backing are also shown. There are other symbols, but the symbols shown in Figure 4.2 are the most widely used.

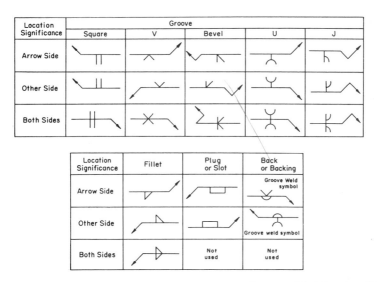

FIGURE 4.2 — *Basic welding symbols and the significance of their locations.*[45]

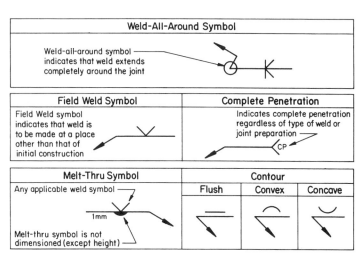

FIGURE 4.3 — *Supplementary symbols used with welding symbols.*[45]

Figure 4.3 shows supplementary symbols used with the welding symbols. Of particular interest to the designer are the complete penetration symbol (CP) in the tail of the arrow and the melt-through symbol placed under the groove symbol. When designing equipment for corrosive service, the CP and melt-through symbols should always be used to help assure that no crevices exist under the groove joints.

Figure 4.4 shows the locations of welding symbol elements. An element is simply a specification, such as the length of the weld, the depth of penetration, the angle of the groove welds, the contour of the weld bead, the root opening, etc., that may be applied to a specific weld. In other words, after the designer has selected the basic joint he requires, he can list the details of how the joint is to be made by adding instructions in the tail of the arrow or at places on the reference line. To accomplish this, the designer may need the assistance of a welding engineer.

To aid the designer in using AWS welding symbols in his designs, examples of their use are exhibited in the following three figures. Figure 4.5 shows a tube-sheet-to-shell welding, a shell seam, and a nozzle-neck-to-shell weld. Figure 4.6 displays a heavy shell seam and joint. Figure 4.7 shows examples of angle-to-plate, shell-to-head, and three t-welds.

The weld locations are shown in the above examples as shaded areas to clarify the use of AWS symbols. However, in actual practice,

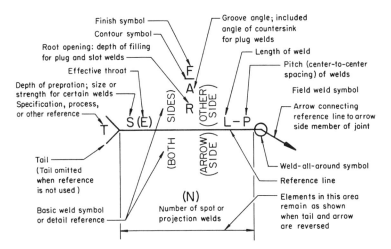

FIGURE 4.4 — Location of elements of a welding symbol.[45]

only the junctions of the joints are shown with the welding symbols pointing towards them. Figure 4.8 exhibits the following four most frequently used joints and how welding symbols are normally used with these joints:

(a) V-groove butt joint
(b) Fillet-welded, corner joint
(c) Square edge joint
(d) Double-fillet T-joint

Other subdivisions of the AWS welding symbol system include flash, seam, spot, protection welding, and even back-gouging. Refer to AWS Specification A2.4-79 ("Welding Symbols") if these other symbols are required.

The Welding Process

The following are various welding processes a designer can specify:

1. Shielded Metal Arc Welding (AWS nomenclature; SMAW), commonly called *stick electrode welding*.

2. Gas Tungsten Arc Welding (GTAW), commonly called *TIG* or *heliarc welding*.

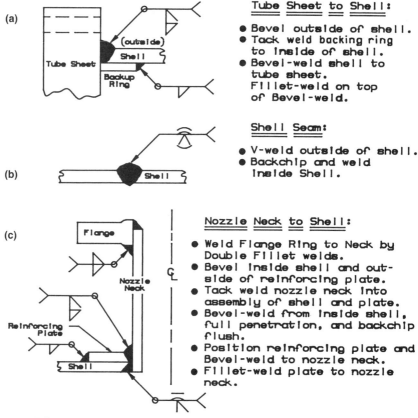

FIGURE 4.5 — *Examples of a (a) tube-sheet-to-shell welding, (b) shell seam, and (c) nozzle-neck-to-shell weld.*

3. Gas Metal Arc Welding (GMAW), commonly called *MIG welding*. This process can be manual, semi-automatic, or automatic welding.
4. Submerged Arc Welding (SMAW), commonly called *subarc welding*. This process can be semi-automatic or automatic welding.
5. Gas Welding. Oxyacetylene is the most common gas welding, although there are other gas combinations.
6. Brazing.

Each process has its individual characteristics; therefore, it is

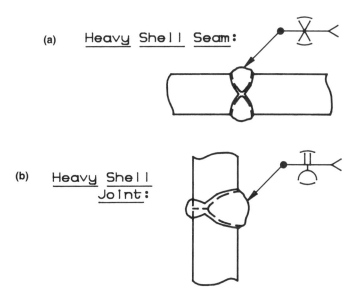

FIGURE 4.6 — *Examples of heavy shell (a) seam and (b) joint.*

important that the designer select the welding carefully so he preserves the material's corrosion resistance. For instance, although the deposit left by the metal-arc process is inherently no less corrosion resistant than the inert gas process, the metal arc process may result in slag inclusions which result from sloppy welding techniques, while the inert gas process does not produce any slag inclusions. As an example, the adverse effect of slag inclusions was pointedly observed in a metal-arc welded stainless steel tank containing an acid. When the tank was emptied, cleaned out, and given a routine inspection, it was noted that the double-butt welded girth weld inside bead had been aggressively attacked, while the adjacent tank wall was relatively unaffected. The cause of the failure was concluded to be very poor workmanship because the slag had not been adequately removed. The remainder of the inside weld was gouged out and rewelded with the inert gas process. After that, no more corrosion problems were reported.

When the fabricator is equally familiar with both metal arc welding and inert gas welding and when practical, inert gas welding should be specified by the designer for corrosive service. Brazing, silver soldering,

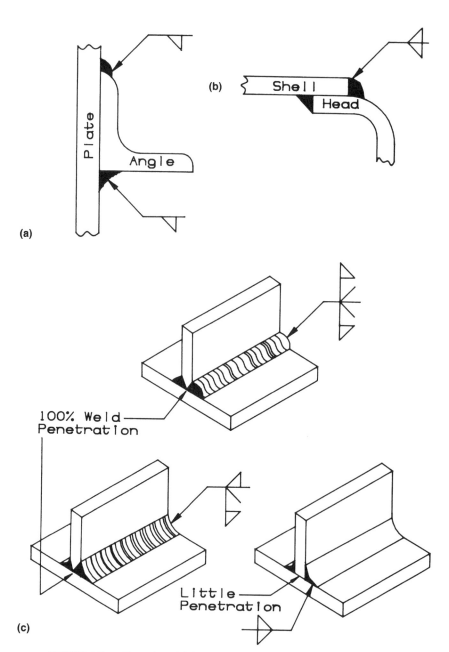

FIGURE 4.7 — *Examples of (a) an angle-to-plate weld, (b) shell-to-head weld, and (c) three t-weld samples.*

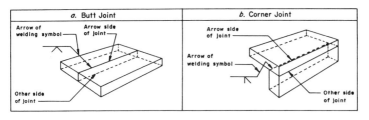
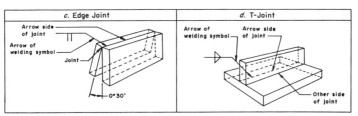

FIGURE 4.8 — *Basic joints; identification of arrow side and other side of joint.*[45]

and soldering should not be specified for corrosive conditions. Exceptions to this rule may exist; however, these joining processes usually entail a different material than the parent metal, which can lead to galvanic corrosion.

Whether or not automatic rather than manual welding can be specified will depend upon the results of the procedure qualification tests discussed below.

Filler Metal

The filler metal, as well as the welding process, should be carefully selected by the designer. Filler metal is very important and should be noted in the specifications, drawings, or both. Many weldments have had limited service lives in corrosive environments because, although the proper materials of construction were specified, the wrong filler metal was used for welding. For instance, many mistakes have been made by welding low carbon stainless steels with higher carbon electrodes, such as welding AISI 304L stainless steel with AISI 308 stainless steel electrodes for use in services that can cause intergranular corrosion. The parent metal perseveres, but the welds fail by corrosion. AISI 308L stainless steel (with a low carbon content) should have been used in such cases.

Qualification Tests

Any proposed welding procedures and welding operators should be qualified according to AWS or American Society of Mechanical Engineers (ASME) Boiler Code Standards. For weldments to be used in critical corrosive service, the designer should specify that all welding operators be qualified or show proof that they have been qualified within the past six months. Ordinarily, welding firms have previously qualified procedures that may be used, provided the necessary qualification affidavits are shown. When a new procedure is proposed, it must, of course, be qualified before welding begins. Figure 4.9 is a photograph of a procedure qualification test in progress for the automatic submerged arc welding of the bottom steel plate for a waste storage tank. It was very important that these welds be sound.

A procedure qualification test is conducted to prove that a proposed weld design and procedure is capable of doing what it is designed to do.

FIGURE 4.9 — Qualification test of submerged arc welding in progress.

After the procedure has been qualified, the operators are qualified to the same procedure. The purpose of the operator qualification test is to prove that the operator can produce a sound weld. The designer should always insist on qualification tests, which can reduce the amount of shoddy welding produced.

Using AWS Specification D10N, ("Qualifications of Welding Procedures and Operators") or Section VIII of the ASME Unfired Pressure Vessel Code, Section IX ("Welding and Brazing Qualifications") is another way a designer can assure good welding.

Wall Dilution

Wall dilution can occur when a less corrosion-resistant metal is welded to the surface of a more corrosion-resistant metal, such as when steel ladder supports are welded to the outside surface of a stainless steel tank. In corrosive applications, the integrity of the metal surface behind the weld must be preserved. Several things can be specified by the designer to accomplish this. He can:

1. Use an over-alloyed welding filler rod to compensate for any wall dilution that would be created by a less corrosion-resistant metal. The use of AISI 309 (25Cr-12Ni) stainless steel filler rod to weld carbon steel to AISI 304 (18Cr-8Ni) stainless steel is an example of this.
2. Specify that the corrosion-resistant metal be thick enough so that the welding arc will not penetrate through the wall's thickness.
3. Specify that both metals be the same. (Because of cost, this may be impractical.)
4. Specify that a transition pad be placed between the two metals. As an example, when joining steel ladder supports to a stainless steel tank, a stainless steel pad is welded with a stainless steel filler rod to the surface of the tank and then the carbon steel is welded to the pad, as shown in Figure 4.10.

In an unusual case, Schedule 10 AISI 309 Cb stainless steel pipeline was required to handle a severely corrosive solution. A carbon steel heating jacket was welded to the stainless pipe. [See Figure 4.11(a).] A filler rod of the same composition as the pipe (AISI 309 Cb stainless steel) was used to join the carbon steel to the pipe. In service, corrosion occurred inside the pipeline where some dilution of the stainless steel wall by the carbon steel from the jacket at the inside surface of the pipe had occurred (this was not burn through). The pipeline was redesigned incorporating transition pads, as shown in

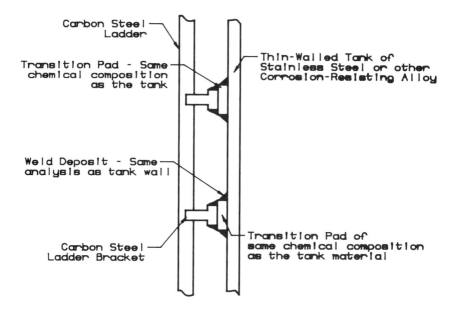

FIGURE 4.10 — *Method of welding carbon steel ladder brackets to thin-walled stainless steel tank.*

Figure 4.11(b). This design required three welds for every one weld in the old design; however, it solved the corrosion problem and was considered worth the extra cost. It should be emphasized that transition pads on jacketed pipelines are usually only required on thin-walled (Schedule 5S) pipe.

Liquid Metal Cracking (LMC)

As discussed in Chapter 3.9, when certain molten metals contact other stressed metals, cracking can result. Of course, the heat of welding provides an ideal environment in which metals can become molten; therefore, the designer must exercise care when selecting combinations of metals to be welded. Zinc, aluminum, and lead in contact with stainless steel can cause cracking; cadmium in contact with titanium can cause cracking; tin and zinc can embrittle aluminum alloys; and tin, lead, and zinc can cause cracking in steels. (See Table 3.7.)

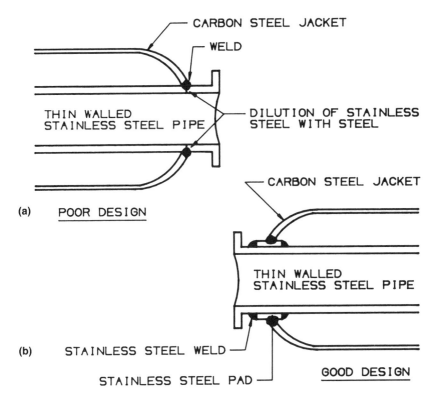

FIGURE 4.11 — *Jacketed pipeline design; (a) poor versus (b) good designs (for thin-walled pipe).*

Zinc causes the most problems because it is used universally as a cathodic protector. For an example[46] of the damage that zinc can cause, see Figure 4.12. The backside of a AISI 316 stainless steel sheet was welded to galvanized angle-iron reinforcements when building a structure used for dryer bag filters. Molten zinc was present during the welding operation and penetrated the stainless steel intergranularly causing liquid metal cracking (LMC). Figure 4.12 shows one of the weld areas on the front side of the stainless steel sheet that had not only become embrittled, but had also become less corrosion resistant. A remedy for this problem would have been to remove all the zinc from the angle iron where the weld to the stainless steel sheet was to be made by cleaning with 20% muriatic acid (or other acid) and thoroughly rinsing and drying before welding. Grinding is often used to remove zinc, but this can leave remnants on the metal surface by smearing.

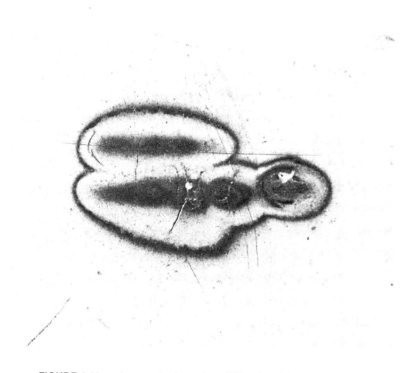

FIGURE 4.12 — *Area on the face of an AISI 316 stainless steel sheet that had been welded underneath to a galvanized angle iron. The zinc had not been removed from the area where the weld was made, thereby causing liquid metal embrittlement and loss of corrosion resistance.*

The Welder's Space

Equipment should be designed so that a welder can position himself to make a sound joint. This may not seem like much of a problem, but it can be very important. In one example, the design required final field butt welds on a cluster of pipes running vertically in parallel. By the time the pipes were ready for field welding, it was discovered that the joints to be welded were down in a narrow slot in the concrete and that the welders could not get close enough to make a full penetration weld. The original drawing clearly indicated that this would happen, but the designer obviously had not considered the space required by the welder. Concrete had to be removed at great expense to allow the welder room to make the welds. Whenever a weld is specified on a drawing, the designer must determine whether or not the welder will have enough space to make an adequate weld. Even the best welder cannot make a sound weld if he cannot get to the joint. If sound welds are not made, corrosion and, of course, mechanical problems may eventually result.

Tube-to-Tube Sheet Welding

As covered previously in Chapter 3.1, tube-to-tube sheet joints provide an area where crevices can create corrosion problems. Rolling the tubes into tube sheets has already been discussed; however, such a procedure never fully eliminates crevices. For this reason, in many applications where corrosion may be severe, the tubes are welded to eliminate crevices on the bonnet side of a heat exchanger or condenser.

The designer has several different weld joints he can specify for joining tubes to tube sheets. Five of these weld joints are exhibited in Figure 4.13.

1. Joint A is the least expensive and hence the most popular joint. It incorporates an edge weld where a filler rod is applied. Sometimes, however, when the inert gas process is used, the tungsten arc is used alone without a filler rod.
2. Joint B is more expensive because the tube sheet hole must be countersunk into the form of a J groove. Filler metal is added.
3. Joint C is another simple joint like Joint A. A fillet weld is applied between the tube sheet hole and a recessed tube. A filler rod must be used.
4. Joint D is not frequently used when the protruding end of the tube is expanded over the tube sheet face and fillet welded with or without filler metal.
5. Joint E is considered the best joint, but is also the most expensive because a groove must be trepanned around each hole. Because the width of each side of the joint is equal, the welding heat is better balanced than the other joints.

Welding tubes to tube sheets will stop any end grain attack or crevice corrosion on the bonnet side of the heat exchanger but the welding heat may open up the crevices on the shell side of the tube sheet that cannot be welded. Therefore, when tubes are to be welded to the tube sheet, the following procedures should be specified:

1. Assure that the hole in the tube sheet and the outside of the tube are clean; otherwise, porous welds may result.

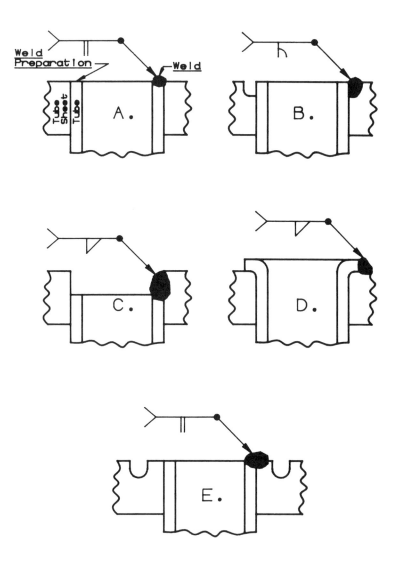

FIGURE 4.13 — *Weld preparation on the left and weld on the right for various tube-tube sheet joints.*

2. Position the tube into the tube sheet, depending on which joint is to be selected. "Stake" the tube by inserting a drive pin and then striking it with a hammer. Staking secures the tube in place with

"line" contact. This action avoids trapping gas between the tube and the tube sheet after the initial roll, which can foster pinholes in the weld.
3. Roll tightly.
4. Weld.
5. Roll tightly again, but not at the weld location to avoid any weld cracking.

Automatic welding of the tube-to-tube sheet joint has been accomplished for many years. Not only can this method be efficient, but it can weld much smaller diameter tubes than can be easily welded manually. Figure 4.14 shows an automatic welder welding 5/8-in.- (16-mm)- diameter Inconel[†] tubes into an Inconel tube sheet. As with manual welding, prototype samples of the expected joint should be requested by the designer.

It is very important that the designer carefully write down the details of his requirement for the tube-to-tube sheet welding in his specifica-

---

[†]Registered trade name.

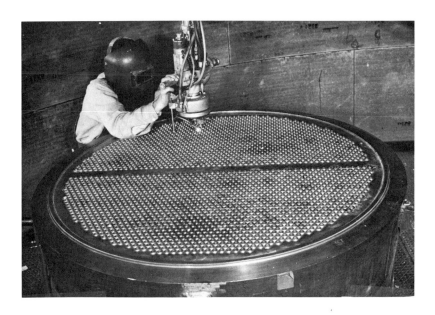

FIGURE 4.14 — A 5/8-in. (0.6-mm) tube sheet welder that eliminates crevices. (Picture by Combustion Engineering.)

tions. As will be addressed more closely in Chapter 5.1, such a specification should be discussed with potential fabricators of the heat exchangers to get their input before issuing the final specification.

Welding Prototypes

For critical applications, the designer should specify that the prototype joints be constructed by the fabricator of the tube-to-tube sheet weld proposed for the actual job. This request should be incorporated in the designer's specifications as a requirement for acceptance. Figure 4.15 shows plan and side views of a sample that was used as a comparison guide during fabrication of a special heat exchanger. An A-type joint was used. Note that the bottom of the tube sheet and the tube joint is very tight and that no overrolling occurs. This particular prototype was approved. Prototypes are sometimes expensive but are worth it for equipment destined for corrosive applications.

Workmanship quality may directly affect the life of a weld in many ways. Undercut, excessive weld reinforcement, underhang or "grapenuts," weld craters, lack of weld penetration, etc., can all serve to collect foreign matter and thus cause corrosion. On jobs intended for aggressive service, actual prototype welded joints are sometimes made to show the welders and inspectors precisely what quality of weld is expected. Requests for prototypes must be part of the designer's purchase specifications.

A case in which a prototype was valuable involved the design of tanks that, while being barged on the Ohio River, were to contain alkylation acid (contaminated sulfuric acid) on the way down the river and strong (uncontaminated) sulfuric acid on the return trip. Absolutely no acid leakage into the river was permitted. State inspections were frequent and unannounced. Because of space limitations, the tanks had to be square and double-fillet welds were required. Prototype specimens were fabricated and sectioned according to the design marked "Poor" in Figure 4.16(a). This design was found to be too hazardous since making an absolutely porosity-free weld is difficult. A pinhole on the inside weld could allow acid to enter the space under the vertical plate section, and this space could then act as a manifold from which the acid could seek out another pinhole on the river side. Consequently, the design was changed to the design marked "Good" in Figure 4.16(b), which incorporates a bevel from both sides insuring a full penetration weld. The chance of one pinhole from the inside coinciding with another from the river side was considered extremely remote. The barges have been in service for many years without any leakage.

Plan View

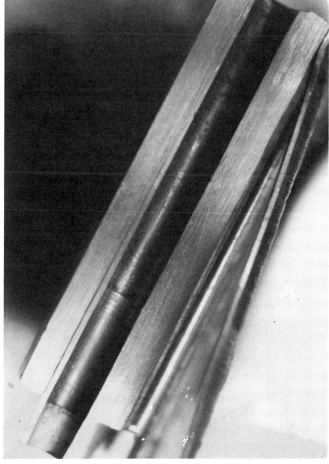

Side view

Note tight joint on shell-side

FIGURE 4.15 — *Sample of tube-to-tube sheet weld for heat exchange; (a) plan and (b) side views.*

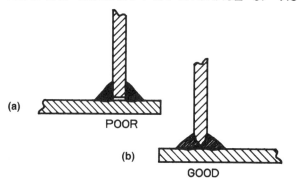

FIGURE 4.16 — Examples of (a) poor and (b) good weld joint designs for barge acid tanks on the Ohio River.

Examples of Excellent Welding

Examples of excellent welding from the Savannah River Plant[1] are shown in Figures 4.17 and 4.18. These weldments were carefully designed using AWS symbols, taking into account the requirement for full penetration welds that do not leave any crevices. The welding procedure was carefully qualified using AWS specifications, and the welding operators were all qualified under AWS rules as well.

The weldments were then expertly fabricated from AISI 304L stainless steel plate and sheet using the GTAW process. The filler rod used was made of AISI 308L stainless steel. The weldments were inspected by dye penetrant and radiographic methods. (See Chapter 6.1.) Such examinations disclosed completely penetrating welds for both the V-groove circumferential body welds and the bevel plus fillet welds for the nozzles.

In summary, when weldments are properly designed and proven welding procedures are followed by qualified operators and the work is properly inspected, sound welds result and consequently, corrosion problems can be substantially reduced.

---

[1] E. I. du Pont de Nemours & Co., Inc., Atomic Energy Division, Aiken, South Carolina.

FIGURE 4.17 — *Example of excellent welding.*

☐ 4.2 Heat Treatment

The thermal history of a metal can have a great effect on its corrosion resistance, as well as mechanical properties. Various heat treatments can be given to metals; therefore, the designer should become aware of the effect of such heat treatments, since occasionally the heat treatment of a metal can circumvent using a more expensive alloy. Heat treatment specifications are very important for the designer.

History

Many years ago, heat treating specifications were also very important. A man's life sometimes depended on his sword. A sword had to maintain its sharpness and be tough at the same time. *Damascus Steel* was the prized material that satisfied both requirements. However, it was very scarce. Because the production of Damascus swords was a closely guarded secret, a great search was conducted to determine a substitute steel that would perform like the Damascus steel. The search

FIGURE 4.18 — *Another example of excellent welding.*

failed because the swords made from the selected steels were either tough and became dull quickly or stayed sharp but broke because of brittleness during the first blow. Years later, it was learned that this marvelous Damascus steel was not a new alloy at all but was simply produced by a new method of heat treatment. The specification of a slower quench from the hardening heat was what made the sword so effective. According to one historian, the heat treatment specification called for a quench in a "bath of urine from a calf collected in the light of the new moon." Figure 4.19 shows an old-time heat treater fulfilling his designer's specification.

Homogenization

As discussed previously, differences in the structure of metals and alloys can lower corrosion resistance. For instance, when metals are produced, differences in the chemical composition from area to area can result. This inhomogeneity can be overcome by a heat treatment, which causes diffusion of atoms in the structure so that a uniform chemical composition results.

FIGURE 4.19 — *Old-time heat treater collecting quench media.*

Annealing may be described as heating metals above a critical temperature range, holding for a certain period of time, and slowly cooling. The annealing temperature will vary with the composition of the metal involved. For instance, the annealing temperature for low carbon steels will vary with carbon content from 1600 to 1700 F (871 to 927 C), while high carbon steels will vary from 1450 to 1500 F (788 to 816 C). The time required to homogenize metals will again vary with the specific metal from hours to several days. Cooling is always slow to ensure a homogeneous structure.

Nonferrous alloys are usually heated to temperatures just below the solidus temperature (just below melting) for annealing.

Relieving Cold Work

Annealing can also relieve residual stresses and cold work in metals. When metals are rolled, forged, or otherwise deformed, the basic structures show highly deformed grains and high internal stresses develop. As discussed previously, these locked-up internal stresses can foster stress corrosion cracking.

Stress Relieving

The stress relieving heat treatment is really a partial anneal. Annealing constitutes three stages, recovery, recrystallization, and grain growth. When it is desired to preserve most of the mechanical properties imparted by cold work, but at the same time (to an extent) maintaining corrosion resistance, a stress relief may be specified. This stress relief duplicates the first stage of an annealing operation.

For low carbon steel, the stress relieving schedule is usually 1100 to 1150 F (593 to 620 C) for 1 hour per in. (25 mm) of thickness, and then the steel is air cooled. When carbon steel is welded, the heat of welding can break up the pearlitic structure of the parent metal and cause the formation of carbides in the heat affected zones. The carbides act as cathodes, and the ferrite areas act as anodes, thereby establishing corrosion cells, as already discussed. This is one of the reasons why carbon steel weldments are often stress relieved (another reason is to maintain dimensional stability).

The following is an example of the use of stress relief. The concentration of a sodium hydroxide solution and its temperature predicates whether or not low carbon steel will stress crack. When there is an indication that cracking will result, a stress relief heat treatment will permit usage without cracking. This applies to steel process equipment, tanks, valves, etc. Table 4.1[47] shows this relationship. As an example of its use, if the concentration of NaOH for a service is 30% and the maximum service temperature is 155 F (68 C), a stress relief should be specified by the designer. If the service temperature for the same concentration of caustic is 130 F (54 C), no stress relief is required.

Solution Heat Treatment

As discussed in Chapter 3.11, it is sometimes necessary to put certain precipitates back into a solid solution to improve corrosion resistance. For instance, unstabilized austenitic stainless steels, when sensitized, may have their corrosion resistance restored if this heat treatment, called *solution heat treatment*, is specified; this treatment involves heating at 1650 to 2000 F (899 to 1093 C) (actual temperature depends on type of stainless steel) for 1 hour per in. (25 mm) of maximum thickness (1 hour minimum) and quenching in water to black heat within 3 minutes. Solution heat treatment places the chromium carbides back into solution.

When either stress relieving or annealing of austenitic stainless steel is thought to be required, the designer should specify only the

TABLE 4.1
Need for Stress Relief of Carbon Steel: Concentration of NaOH vs Service Temperatures[1]

Concentration of NaOH (%)	Service Temperature		
	No Stress Relief Required Ambient to: (F)	Stress Relief[2] Required: (plus nickel trim on valves) (F)	Use Nickel Alloys (F)
10	180	over 180 to 220	over 220
20	160	over 160 to 230	over 230
30	140	over 140 to 220	over 220
40	125	over 125 to 180	over 180
50	120	over 120 to 170	over 170

[1] Table based on findings of NACE T-5A Committee.
[2] Stress relief heat treatment: Hold at 1100 to 1150 F for 1 hour/in. of maximum thickness (minimum time was 1 hour).

solution heat treatment. If the equipment involved has a geometry that will not allow it to take the water quench required by this heat treatment without warping, the designer has two options; he can

1. Decide whether the heat treatment is really necessary, or
2. Change to using a low carbon or stabilized stainless steel that does not require a heat treatment to preserve corrosion resistance.

Stabilizing Heat Treatment

A stabilizing heat treatment, which is occasionally used for treating the higher carbon stabilized stainless steels, involves maintaining the temperature between 1500 and 1650 F (816 and 899 C) to allow sluggish titanium and/or columbium carbides to be prompted into precipitating, rather than having chromium carbides precipitated, which may lower corrosion resistance. This heat treatment should be specified with care because a hazardous condition can be encountered, as follows.

Certain austenitic stainless steel alloys are susceptible to the formation of a hard, brittle compound of iron and chromium called *sigma phase* when they are exposed to a temperature range of about 1300 to 1650 F (704 to 899 C). Fully austenitic stainless steels (i.e., containing no ferrite) usually do not develop this phase. However, when ferrite is present, as in AISI 309 stainless steel, sigma phase can develop.

One reported case[17] involved heat treating a vessel made of AISI 309 Cb stainless steel. After fabrication, this vessel was given a stabilizing heat treatment that satisfied the requirements of the ASME pressure Vessel Code. However, the problem was that sigma phase was formed. The AISI 309 Cb vessel in the example was unsatisfactory in service for this reason. If the solution heat treatment had been specified rather than the so-called "stabilizing" heat treatment, which was in the sigma-forming temperature range, the vessel would have remained ductile and satisfactory for service.

What the Designer Should Know About Heat Treatments

The designer should know about the various heat treatments available for the particular metal or alloy he is planning to use. It is best to consult with a metallurgist to determine the actual need for heat treatment and, if required, what the schedule should be.

The designer should specify the full heat treatment schedule required. A notation of just "anneal, stress relieve, solution heat treatment," etc., is not adequate. An example of a proper notation is a stress relief schedule specified for large waste storage tanks, 60 ft (18 m) in diameter and 35 ft (11 m) high, to obviate stress corrosion cracking and, at the same time, to avoid warping the large tank. The following is an example of a heat treatment schedule:

1. Heat to 600 F (315 C).
2. Above 600 F (315 C), heating is not to exceed 100 F (38 C)/hour. During this period, the temperature gradient is not to exceed 125 F (52 C) in any 15-ft (5-m) interval and then there should not be a greater variation than 200 F (93 C) between the lowest and highest temperature points in the vessel.
3. The temperature is to be held at a minimum of 1100 F (593 C) for a period of at least 1 hour.
4. The rate of cooling should not be greater than 125 F (52 C) per hour. During this period, the greatest variation between the highest and the lowest temperature in the vessel is not to exceed 200 F (93 C).
5. Below 600 F (315 C), no restriction on the cooling rate is required.

The above heat treatment schedule was successful since very little, if any, warping occurred. More importantly, no stress cracking has occurred since in any of these heat treated vessels. The heat treatment schedule is not usually as detailed as the schedule described above. For instance, "Heat slowly to a temperature of 1100 to 1200 F (598 to

648 C) and hold for 1 hour per in. of thickness, then furnace cool to ambient temperature" could be used for stress relieving a piece of production equipment not susceptible to warping.

No matter which alloy is used, the complete heat treatment should be clearly defined in the designer's specification. However, some engineers argue that such a treatment is up to the heat treater's discretion, and all the designer needs to do is to specify the end result, such as the desired hardness of the part or equipment. Consulting with the heat treater first is a prudent step, but the designer should specify the heat treatment schedule agreed upon.

For example, tool steel blocks that were to be used as important parts in a piece of equipment were sent to a heat treater with the specification that the blocks "are to be heat treated to a Rockwell C hardness of 63." After the parts were quenched from the hardening temperature, the heat treater found that the parts were already at the required hardness so he did not bother to temper the parts (like he should have) because of fear that the parts would become too soft. Consequently, the blocks were so brittle that they failed immediately when used. No recourse was expected from the heat treater because tempering had not been specified. The designer should always specify the complete heat treatment schedule, including temperature, time at heat, quenching medium, quench temperature, and the tempering temperature (when a temper is required). Such a specification can also be used by the inspector later to assure that the required heat treatment had been accomplished.

☐ 4.3 Temperature

Effect of Temperature on Corrosion

Similar to most chemical reactions, the intensity of corrosion attack generally increases as the temperature rises. One rule-of-thumb is that the reaction rate (kinetics) doubles with each 10 C rise in temperature. The designer must consider this effect when designing to prevent corrosion. Perusal of the Corrosion Data Survey Charts (See Table 2.13) discloses how most metals experience higher corrosion rates with increasing temperatures. For this reason, temperatures in general should be kept as low as is practical.

Materials of construction vary widely in their abilities to withstand temperature. For instance, low carbon steel may be used in service up to about 850 F (426 C), while austenitic stainless steel may be used up to temperatures around 1600 F (871 C). Alloys with chromium contents of 20% or more may be used at higher temperatures 1900 F (1038 C) and, in certain instances, above that temperature.

Resistivities of soils and waters normally decrease with rising temperatures, thus promoting higher corrosion rates. Also, the resistivity of seawater, which will be discussed later in this chapter, can be markedly lowered by higher temperatures and hence accelerated corrosion can result.

Hot Wall Effect

Certain design features exist that can foster higher temperatures, which may not be fully realized by the designer. For instance, the heat transfer surfaces of process equipment may be susceptible to higher temperatures and hence higher corrosion rates. As an example, the surface temperature of heating coils or of steam jackets is appreciably higher than the maximum temperature of the fluid being heated in order to get a transfer of heat. This is termed the *hot wall effect*. The designer must select appropriate materials of construction for these surfaces based on the higher temperature, not on the solution temperature.

As an example, suppose a steam coil is to heat a solution to 180 F (82 C). In determining the material of construction for this coil, suppose specimens of candidate materials were suspended in a simulated process solution operating at 180 F (82 C) and, on the basis of corrosion rates determined by this test, a materials decision was made. This decision could be wrong and pipe made of this material could fail rapidly because the hot wall effect was not considered. For the coil wall to transfer heat energy to the solution to elevate it to 180 F (82 C), the wall itself may be elevated to a temperature above 210 F (99 C). This higher temperature could make a significant difference in the ability of the selected materials to withstand the corrosive conditions. Many materials mistakes have occurred because this phenomena was not considered.

A factor that can influence the temperature of heat transfer surfaces is the method of heat transfer, which can occur in several ways, including convection, film boiling, and nucleate boiling. These last two methods particularly can cause high surface temperatures that lead to excessive corrosion. When film boiling occurs, the heat transfer has to take place through a poorly conducting film barrier, thereby raising metal temperatures. When heat transfer occurs through a vapor formation at local areas, it is called *nucleate boiling*, which results in high temperatures at distinct spots. Not only is the temperature high in nucleate boiling but high concentrations of dissolved solids occur at these areas, which leads to excessive pitting. Heating by convection causes the least amount of problems.

What the Designer Can Do to Reduce High-Temperature Problems

The designer can partially avoid these high-temperature problems by adhering to the following guidelines:

Provide good circulation of the fluid to foster heating by convection. This type of transfer will elevate the temperature the least, as compared to other types of heat transfer.

Specify periodic cleanings. Make sure that all heat transfer surfaces remain as clean as possible by specifying periodic cleanings. Clean surfaces can reduce nucleate boiling.

Suppress boiling. When practical, specify a back pressure for the equipment so that any boiling will be suppressed.

Make sure all heat transfer surfaces are always submerged. Have equipment used for heating corrosive solutions, such as coils and reboilers, designed so that the heat transfer surfaces are always submerged. When only partially submerged, the surfaces will not only overheat, but the corrosive fluids can concentrate as well. The designer can assure complete submergence by designing fluid-vented tube sheets or effluent nozzles located at a much higher level than the inlet nozzle and the heating elements. (See Figures 3.65 and 3.66.)

Simulate the hot wall effect using test methods. The designer should take advantage of the methods described in Chapter 2 when selecting materials of construction for heat transfer surfaces. Heat transfer calculations can also be made to estimate the surface temperature of the heating elements. The actual surface temperature, of course, will depend on, among other things, the heat flux and hence the rapidity with which the solution is heated.

Avoiding High Local Heat

The designer should avoid, as much as possible, any areas of high local temperature in his designs. For instance, heating elements should be located away from vessel walls. (See Figure 4.20, which shows good and poor designs.) Proper location not only affords more efficient heating but also avoids overheating of vessel walls adjacent to the heaters.

Direct flame impingement should be avoided. If it is impractical, select a material that can resist the high local heat. For example, in an arms plant, gas-fired lead melting pots for the shotting towers had to be replaced frequently because of side-wall failure. An examination of the carbon steel pots revealed that the failures were occurring from the

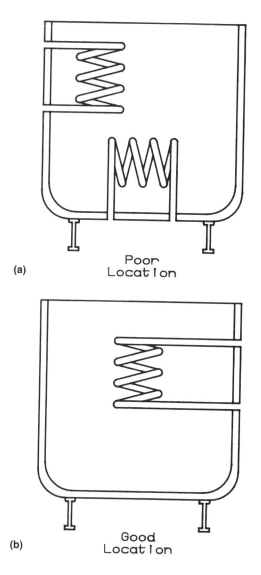

FIGURE 4.20 — (a) Poor and (b) good locations for tank heating coils.

outside rather than from the inside. New pots were designed with 20% clad AISI 304 stainless carbon steel as the material of construction with stainless steel located on the outside of the pots. The pots have held up under the high local heat and have given good service without having to be replaced for many years.

Cold Areas

Cold areas have caused almost as many corrosion problems in industry as hot areas. As an example, boiler flues unexpectantly failed in certain areas and not in others, even after the high-temperature environment had been carefully considered when the materials were selected. The hot flue gases alone did not corrode the carbon steel, but at local areas where the temperature inadvertently had dropped enough for the sulfur oxides (SO_2, SO_3) in the flue gas to condense to sulfuric and sulfurous acids, excessive corrosion did occur. Acid concentration in the condensate was found to be around 30%. Carbon steel could not be expected to last very long under these conditions. This type of attack may be expected when sulfur-containing fuels are burned. The presence of even 5 parts per million (ppm) of sulfur trioxide may raise the dew point by as much as 100 F (38 C),[48] thereby causing condensation of acid at a much higher temperature than would be expected in the absence of SO_3.

In a similar incident, it was recognized that cold spots could cause corrosion problems; however, because of the improper application of thermal insulation,[49] corrosion occurred anyway. Figure 4.21 displays an insulated reactor containing hot gases. On the left side of the figure, a steel support was left bare. The uninsulated steel support lost heat to the extent that a cold spot was formed. At an area directly above the steel support in the inside of the reactor, condensate formed and corrosion occurred. Consequently, extensive welding repairs had to be made. The proper design, as shown on the right side of Figure 4.21, is to insulate the steel supports so that no cold areas would be formed.

☐ 4.4 Free Drainage

When water becomes quiescent, concentration cells can be formed that can, as mentioned previously, cause aggressive corrosion. Anything interrupting or stopping the flow of water can be instrumental in establishing corrosion cells. For this reason, the designer should eliminate blind corners and recesses from his designs in which solids and liquids can accumulate. He can accomplish this by providing fillets

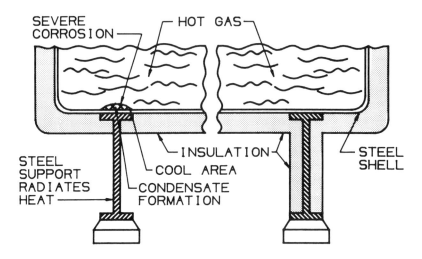

FIGURE 4.21 — *Design of insulated reactor containing hot gases; (left) cold spot causing condensate to form and corrode the steal reactor; (right) support leg insulated to obviate corrosion.*

(by welding or forming), instead of sharp corners, and by providing free drainage.

1. To assure free drainage, structural members should be arranged according to Figure 4.22. Angles, channels, and other structural shapes located outdoors or where they can become wet should be positioned so that they cannot trap water. (See the example marked "Best" in Figure 4.22.) When positioning of structural members is not practical, drainage holes should be provided so that if the structurals fill with water, they will retain the water for only a short time. (See the example marked "Better" in Figure 4.22.) Angles should be closed by welding to avoid water retention, and angles should be welded all around to avoid areas that will trap water.

2. An example of what corrosion problems can be caused when free drainage is not provided involves a major oil refinery,[38] in which about 65 miles (104.60 kilometers) of aluminum instrument air line tubing was installed outdoors to serve a catalytic

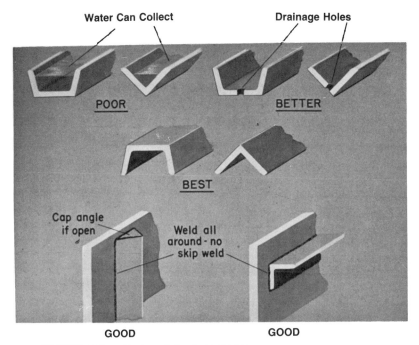

FIGURE 4.22 — *Design of structural shapes.*

cracking unit. The tubing was supported in steel channels with the flanges up. Apparently, no drainage holes were provided. Rainwater held in the channel, because the flanges of the channel were up, resulted in extensive pitting and perforation of the tubes. Miles of tubes had to be replaced.

3. Tanks and process equipment should also be designed, keeping proper drainage in mind. Figure 4.23 shows poor and good methods, and the best method for providing this drainage. Any design where a "heel" of solution can inadvertently remain in a tank after it has been supposedly emptied can lead to pitting. This is harmful because the corrosion might not be noticed until a perforation through a side wall of the bottom occurs. This can be disastrous if the solution handled is hazardous or is toxic.

When designing a tank, if the designer draws a straight line to indicate a flat bottom, he, no doubt, will not get a flat bottom, especially in a large tank, unless he states exactly what flatness tolerance is required. A tolerance noted as "flat within 1/4 in./ft (6 mm/0.3 m) or other similar notation should be specified on the drawing or the specification. One so-called "flat-bottomed" steel tank, 40 ft (12 m) in diameter, was used for storing 66° Baume

(93%) sulfuric acid. The tank started to leak, causing severe pollution problems. Upon examination, the bottom was found to contain ripples 1-in. (25-mm) deep. When the tank stood "empty," strong acid retained in the ripples was diluted by moisture from the air to below 78%, and corrosion of the steel rapidly occurred.

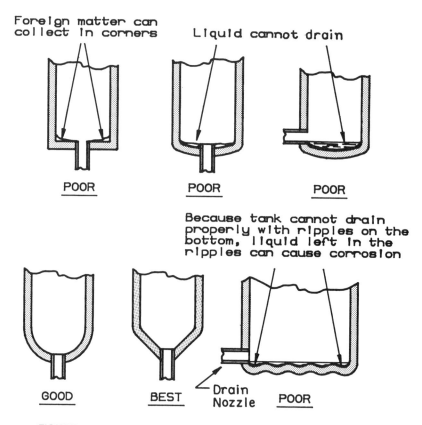

FIGURE 4.23 — *Poor, good, and best tank drainage designs.*

4. When a flat-bottomed tank is to be supported on a concrete pad, crowning of the pad can assure free drainage from under the tank bottom. (See Figure 4.24.) Making the pad slightly smaller in diameter than the tank can minimize the problem of water and other spilled liquids from being drawn under the tank.

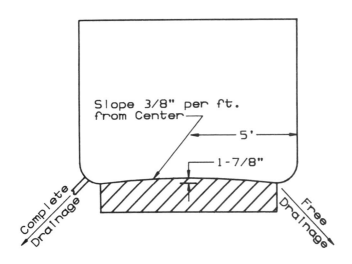

FIGURE 4.24 — Use of crowned concrete pad to assure free drainage of tank and under tank contents.

5. Although protective coatings are not normally associated with hindering free drainage, porous coatings *can* obstruct free drainage. For example, at a chemical plant in Niagara Falls, steel columns supporting the roof of a process building were supposedly protected by a bitumastic coating from hydrogen chloride in the building atmosphere. A fork lift truck inadvertently bumped up against one of these columns and surprisingly the column moved about 6 in. (152 mm); the column had fractured at a corroded area. Examination disclosed that the columns, although appearing sound from the outside, had been seriously corroded by the moist atmosphere. Opening up the coating showed a liquid inside contacting the steel column, which had a very low pH level, indicating the presence of hydrochloric acid. Uncoated steel in the same building had not been significantly corroded. Apparently, the moisture drained off rapidly. On the other hand, the reason for the deterioration of the columns was that the porous coating trapped the acidified moisture, enabling the corrosion to occur.

6. The proper sloping of tubing and pipelines to permit thorough drainage can reduce corrosion, particularly if the process is subject to long shutdowns. If such sloping is not practical, the

designer should specify equipment and connections to blow out the pipelines before long shutdown periods.

7. For lined equipment and vessels that periodically pour out corrosive solutions, a lined overhanging lip should be specified to avoid corrosion of the equipment's side walls by liquid flowing down.

8. Concrete plinths should be specified for drainage of the support legs of equipment. The plinths ensure the complete drainage of any fluids that may inadvertently contact the support legs, such as spills, cleanup water, etc. On many occasions, excessive corrosion at the bottom of steel support legs has been observed, while the upper part of the legs were in good condition. Invariably, this problem has been caused by legs standing in water (and other fluids) for long periods of time. Especially important is that the designer incorporate plinths in his designs for equipment to be located outdoors. (See Figure 4.25.)

☐ 4.5 Environmental Corrosion

Atmospheric Corrosion Problems

Although the designer is primarily concerned with the corrosion of metals in process solutions, the effect of the outside atmosphere on structural metals and alloys is equally as important. The geographical location and variable concentrations of soot, flyash, sulfur compounds, suspended solid particles, carbon dioxide, and chlorine compounds coupled with moisture and oxygen in the air can greatly affect the corrosiveness of the atmosphere.

Because of the wide variations in the corrosiveness of the atmosphere, the designer should carefully consider where a new plant is to be located before specifying structural materials, outside tanks, process equipment, etc. Marine, rural, municipal, and industrial atmospheres each have a set of corrosive conditions.

Sulfur compounds formed by burning high-sulfur-bearing fuels are the most corrosive in air, except of course moisture and oxygen. When these compounds come into contact with the atmosphere, they form sulfur dioxide and hydrogen sulfide. When these gases contact moist air, they are hydrolyzed into sulfuric acid. Being gases, they can drift far from their sources. Contact of the acid with steel can cause corrosion

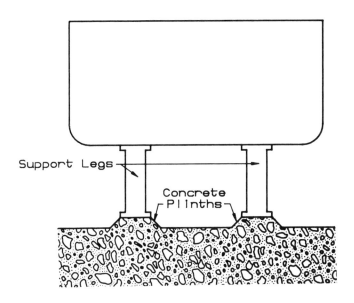

FIGURE 4.25 — *Concrete plinths to ensure complete drainage to avoid corrosion of steel support legs.*

and ferrous and ferric sulfates are formed. These compounds are hydroscopic and by absorbing moisture, they can maintain corrosion over a long period of time.

Any new anticipated plant location should also be carefully studied by the designer to determine the severity of atmospheric corrosion to expect. For instance, high SO_2 content in the atmosphere can be very aggressive and is sometimes found in heavily industrialized areas. For instance, in some sections of Northern New Jersey, the SO_2 content in the atmosphere is so high that it has to be monitored by the state using aircraft.

In marine atmospheres, because of salt in the air, corrosion can be severe, depending upon the distance from the shore. If within one mile of the shore, corrosion can be significant. The corrosiveness of the atmosphere lessens with farther and farther distances from the shore. After about six or seven miles (10 or 11 kilometers), the corrosion by chlorine compounds is insignificant.

Rural and municipal atmospheres generally cause the least corrosion as compared to marine or industrial atmospheres, because no strong chemical contaminants are present other than moisture, oxygen, and carbon dioxide.

What the Designer Can Do to Reduce Atmospheric Corrosion

When atmospheric corrosion may be a problem, a designer can use the following guidelines:

Test candidate materials at the plant site. When locating a new plant in an area where the corrosiveness of the atmosphere is unknown, the designer should arrange to have candidate materials tested at the plant site. The procedure to follow is the same as the procedure outlined in Chapter 2.

Consider industrial neighbors. During plant layout, the designer should consider the industrial neighbors that will be close to the anticipated plant. For example, a small plant was to be built downwind of a Mannheim furnace, which produces muriatic acid. Specimens of AISI 410 stainless steel were installed at the site. These specimens corroded almost as rapidly as carbon steel. Fortunately, the test was made before the plant was built and relocation of the plant was still possible.

Be careful about water-spray-producing equipment. Large pieces of equipment that introduce substantial amounts of water spray into the surrounding atmosphere can also intensify corrosion. Such equipment includes water cooling and quench towers. The proper positioning of this equipment relative to other equipment is vitally important. When possible, the designer should locate such equipment as far away as possible and leeward to other structures.

Consider other methods of reducing atmospheric corrosion. All steel equipment should be coated with a paint or protective coating that is resistant to the particular environment confronted. (See Chapter 6.6) The designer should specify that any vessels and structures be painted before assemblage and delivery to the field because the surface preparation is vital to the integrity of the final coating. Trying to paint a tank in the field after it has rusted is expensive because of the required surface preparation. In Japan and in several steel mills in the US, steel sheet can be procured that has been prime coated at the mill, which can save money in the long run.

Not only is it necessary to assure that the proper protective coating is installed over properly prepared surfaces, it is also necessary that the coating is maintained over the years. The Du Pont Company's Automotive Products Division offers a protective coating service that not only applies protective coatings, but also maintains them over the years on a guaranteed basis. These services must be specified in the original design, not years afterwards when the structures have already been corroded.

The designer can also specify that structurals in critical areas be galvanized prior to assemblage. Such structurals may be used as is or with protective coatings. The zinc acts as a cathodic protector, as discussed in Chapter 3.2. Surface conversion pretreatments with phosphoric or chromic acids where phosphates or chromates are deposited on the metal can greatly extend the useful life of protective coatings. Other metallic coatings for steel parts to be used outdoors can be specified and applied by spraying or electrodepositing. These coatings include aluminum, lead, copper, cadmium, chromium, and nickel.

Specify cost-efficient, corrosion-resistant materials. The designer should specify corrosion-resistant materials which are consistent with the cost considerations. When solid alloys are too expensive, steel clad with the alloy is much less costly when the required thickness is over about 3/8 in. (10 mm). The welding of clad materials does not present any problems when the proper welding specifications are being followed.

Avoid any corrosion-causing factors. The designer should avoid all factors that will cause water condensation, or the collection of corrosion products, or dust and debris in outdoor exposures. This includes many subjects already discussed, such as rainwater retention in crevices, incomplete welds, avoidance of sharp corners, etc.

☐ 4.6 Fresh Water

Fresh Water Corrosion Problems

The aggressiveness of so-called "fresh" water depends upon many factors, such as the amount of dissolved gases and solids, oxygen content, temperature, the water velocity, and the ability to form protective scales or films. *Hard* waters are generally less corrosive than *soft* waters because of the precipitation of magnesium and calcium carbonates on the surface of metal parts, which acts as a protective coating and lessens corrosive attack. Many fresh waters contain silicates. These compounds are beneficial because they also form protective films. As a matter of fact, silicates are often added to waters to act as corrosion inhibitors.

Acidic mine waters can be extremely corrosive, as can water from deep wells containing hydrogen sulfide. When water smells like rotten eggs, aggressive corrosive action can be expected.

Decomposed animal and vegetable matter in water alone can lower the pH level to such an extent that the water will rapidly attack many metals. On the other hand, waters with a low acid content (above a pH value of 7) will be alkaline in nature and corrosion will be limited.

Water in rivers that become polluted by industrial plants, municipal sewage systems, etc., can be exceedingly corrosive. Waste introduced into rivers can react with the mineral content or organic material in the river, thereby producing such gases as carbon dioxide, sulfur dioxide, and hydrogen sulfide. In turn, these gases can form acids with water, which can make the water very aggressive when used by industrial plants.

Dissolved oxygen is the most corrosive component in water. Depending on the amount of oxygen, water can be very corrosive or very mild. As an example, a heating system that had been operating for 45 years at a paint plant was examined. Because of safety considerations, a particular vessel could not be fired directly, and therefore a coil was designed to wrap around the vessel. The coil led to another room where an oil fire burner heated the water inside the tube. The hot water was then circulated back to the vessel, which supplied the necessary process heating. In 45 years of service, little or no sign of corrosive attack on the low carbon steel coil was observed. The reason was that the first few times the water was heated, the dissolved oxygen was boiled out. Over the years, no make-up water needed to be added. Because dissolved oxygen was not present, little corrosion occurred.

What the Designer Can Do to Reduce Fresh Water Corrosion Problems

Below are guidelines a designer can follow to reduce fresh water corrosion problems.

Analyze the water to be used. Have the water the plant intends to use analyzed. When appropriate, consult a water expert to interpret the results.

Do not specify galvanized steel for hot water service. Waters containing carbon dioxide in combination with oxygen can be very aggressive to galvanized steel when the water is hot [above 165 F (74 C)]. Although galvanized steel is useful for handling cold water, the designer should not specify it for hot water service. When certain chemicals are present, it has been reported that under heated conditions, the zinc becomes cathodic to the steel and accelerates attack.

Expose candidate materials to aggressive water. If tests show the water to be aggressive, expose candidate corrosion specimens to the water. Determine the corrosion rate for each specimen before the materials are selected. In addition, the Langelier Index of the water can be determined. This index is useful in predicting the scaling behavior of the water.

Determine if microbiologically influenced corrosion is a problem by testing the water. If so, take appropriate action, as noted in Chapter 3.7.

Survey the area for potential sources of pollution. Check on the type of industrial plants or other polluting concerns that may be upstream of the contemplated plant. Try to determine if any organizations are anticipating locating a future plant on the same river that might contaminate the water. This knowledge may mean the difference between specifying stainless steel instead of carbon steel pipelines.

Seawater

Because of the interest in the corrosiveness of seawater dealing with desalting operations throughout the world, this subject is being emphasized here.

Seawater variance. The corrosiveness of seawater varies greatly. Several examples will illustrate this point.

Example 1 —Data[50] from a joint test program of the Los Angeles Department of Water and Power and the International Nickel Co.[2] disclosed that an 1/8-in.- (3-mm)-thick AISI 304 stainless steel specimen was perforated in a 3-year test specimen. The specimen showed crevice corrosion beneath barnacles at sheared edges. In another test,[50] medium carbon steel corroded at a rate of 176 mils per year (mpy), high-tensile steel at 220 mpy, and cast iron at 600 mpy in a high-velocity seawater impingement test at 13.5 ft/s [247 meters/minute (m/min)].

---

[2]Formerly the International Nickel Co. is now Inco Alloys International, Huntington Beach, West Virginia.

Example 2 — In another example, a 280-year-old French cannon made of iron, which had been recovered from seawater at a great depth, was found to be practically untouched after the marine growth had been removed. Little or no corrosion was found because very little oxygen was present because of the depth of the ocean.

Oxygen content. The oxygen content of seawater is very critical. For instance, if all the oxygen were removed from seawater, steel and copper would show much better corrosion resistance. The oxygen content in seawater varies widely with depth. For example, the oxygen level in seawater off of California's Coast has been determined[50] to drop steadily from 6 milliliters/liter (mL/L) at the surface to less than 0.5 mL/L at a 2000-ft (610-m) depth. The oxygen content can vary widely, depending on wave action, depth, and temperature.

Total dissolved solids. The word "seawater" does not refer to a single solution or electrolyte. *Seawater* refers to waters that contain total dissolved solids (TDS) varying from 8000 ppm in the Baltic Sea to 60,000 ppm in the Arabian Gulf bay areas.

Biological activity in seawater will also vary widely. Seawater is a complex mixture of many salts, living matter, silt, dissolved gases, and decaying organic matter. Depending upon the balance of these factors, seawater can be rather noncorrosive or exceedingly corrosive.

Because of the variable nature of seawater, it is very important for the designer to carefully select the materials of construction. The following sections are a general guide that a designer may use for selecting materials of construction.

Piping in seawater (low pressure). Although low carbon steel piping can be used in seawater service because corrosion will not be high, the products of corrosion can plug up reverse-osmosis permeators and drastically affect yields downward. Actually, the installed costs of plastic piping is about the same as those for carbon steel pipe. (Comparative costs will vary, depending on location.) Many types of plastic piping are available, including fiber-reinforced polyester (FRP) and polyvinylchloride (PVC). Plastic piping can give excellent service in seawater applications if the designer takes the following precautions:

1. Do not pressurize over the manufacturer's recommendations.
2. When seawater is laden with sand or silt, the flow rate in the pipe should not exceed 10 ft/s (183 m/min),[51] or excessive erosion can occur.
3. Screens should be specified for intakes to reduce the amount of sand and silt that can be entrained.

4. Keep piping turns at a minimum, and when necessary, use piping turns with wide sweeping designs.
5. The type of pipe used should not contain graphite if stainless steel is in the same system, since particles may become dislodged and settle on the austenitic stainless steel surfaces, thus causing galvanic corrosion. (The graphite will become the cathode and the stainless steel, the anode.)
6. Because of the comparatively weak shear strength of plastics compared with metals, care must be exercised in laying the pipe in the ground and especially under sand to make sure that there is solid fill under the length of pipe. If shifts occur in the sand or earth, or voids are left under the pipe originally, the pipe can fail by shearing. For this reason, fiber-reinforced plastic piping is better than solid pipe because of its increased strength. When large diameter pipe is required but is not available in plastic form, concrete-lined steel pipe should be considered.

High-pressure service in seawater. In some desalting operations, such as a reverse osmosis systems like Du Pont's Permasep,[†] pressures of up to 1000 psig are required too high to permit the use of plastic piping. Instead, plastic- or rubber-lined steel pipe can be used. AISI 316L stainless steel could also be used if it is completely flushed with fresh water after shutdown.

1. *Seawater pumps.* Naval vessels[52] use the following materials of construction for seawater pumps: G-bronze for the body and nickel-copper Alloy 400 (Monel[†]) for the impeller and pump shaft. Other pump materials widely used in coastal power plants are cast austenitic steel body (CF8M: AISI 316 stainless steel castings) and AISI 316 impeller with a nickel-copper Alloy 400a or K-500 shaft.
2. *Stainless steel in seawater.* The successful use of austenitic stainless steel pumps, pipe, and other operating equipment is predicated on always keeping the velocity up to a *minimum* of 5 ft/s (91 m/min). When equipment must be shut down, it must be drained and thoroughly flushed with fresh water immediately to assure that no seawater will lie undisturbed on a stainless surface, thus causing excessive corrosion. (Review the example of seawater corrosion in Chapter 3.4.)

[†]Registered trade name.

3. *Other precautions.* The designer should follow other precautions for design mentioned previously in this book. Crevices and dissimilar metals are particularly sensitive to severe corrosion in seawater.

Corrosion in the Soil

Corrosivity of soil varies widely depending on acidity, water, and oxygen contents much like the variance described under fresh water and seawater. When the water content is very low, corrosion is unlikely to occur on buried structures and pipe. As the water content increases, the corrosivity generally increases.

In an example, AISI 304 stainless steel piping was buried in the ground during the building of a chemical plant. When the plant was completed, a valve was turned on supposedly filling the stainless steel pipeline. However, nothing came out. Investigation showed that the pipeline was perforated with many holes. The cause of the corrosion was determined to be a very corrosive soil coupled with the presence of stray electrical currents.

In another example in Bridgeport, Connecticut, the representative of a local establishment served on a municipal pollution board. This organization was trying valiantly to clean up the river and bay. One of the major problems was oil contamination. Investigators finally tracked down one of the culprits, which turned out to be the representative's company. A large buried steel oil tank was subsequently uncovered and found to be full of holes. When the tank had been buried, it had had a protective coating that was practically nonexistent at the time of the examination by this author. A cavity in the ground surrounding the tank contained more than 10,000 gallons (37,850 liters) of oil. Apparently, the tank had leaked into the river for many years. The soil had been very corrosive to steel. The plant was advised to replace the steel tank with an FRP tank. It was further advised that the new tank be located 6 ft (2 m) above ground so that any leaks that might occur would be readily detected.

What the Designer Can Do to Reduce Corrosion in the Soil

There are several actions a designer can take to reduce corrosion in the soil, such as:

Have the soil analyzed and its conductivity determined. Such tests can usually give a good indication of the corrosiveness of the soil.

Remember that buried steel tanks, no matter how they are coated, have a finite life. As mentioned previously, this is estimated to be around 15 years. However, the life of a buried tank could be much shorter, depending on the condition of the soil; therefore, buried tanks should be guarded by cathodic protection. Without this protection, it only takes one pin hole in a buried tank to start leaks. There are several options.[1]

—Do not bury the steel tank, or
—Use FRP tanks which are very resistant to soil action.

Use protective coverings or coatings. All buried pipelines, whether carbon steel, stainless steel, or other metals, should be specified to be protected by some kind of covering or protective coating. Carbon steel is popular as a pipeline material because of its low cost and strength, but it must be protected. Depending on the aggressiveness of the soil,

—Specify an application of coal tar enamel using:
American Water Works Association Standard or
AWWA C-203 for Coal Tar Enamel.
—Specify a protection coating plus a tight wrapping of plastic, such as polyester, polyvinyl chloride (PVC), or polyethylene.
—Specify other types of special organic and fibrous coverings.

Specify plastic piping. When mechanical properties and process conditions permit, specify FRP, epoxy fiberglass, PVC, or other plastic piping.

☐ 4.7 Automotive Corrosion

Over the years, the corrosion of automobiles has presented a challenge to the automotive designer. However, great strides have been made in improving this situation because the prevention of corrosion is one of the important considerations in designing automotive products today. The automotive designers' primary corrosion problems include general corrosion, crevice corrosion, drainage, and galvanic corrosion. They use inhibitors, chemical surface treatments, protective coatings, plating, rust preventive oils, and other measures to help prevent corrosion. Galvanized steel is a preventive measure used in automobile frames. However, the foremost preventive measure concerns design details.

---

[1]Also see NACE Standards RP0169-83 *Control of External Corrosion on Underground or Submerged Piping Systems* and RP0285-85 *Control of External Corrosion on Metallic Buried, Partially Buried, or Submerged Liquid Storage Systems.*

Basis for Design

The Society of Automotive Engineers (SAE) publication SAB J447a cites the following basis for designing a car[53] to improve corrosion resistance:

1. The best corrosion protection is obtained by avoiding, or at least minimizing, exposure of metal surfaces to the continued service environments of water, salt, and other road contamination. The duration of wetness of each exposure determines the degree of protection or the degree of corrosion resistance required for each component.

2. It is often more economical to provide good corrosion prevention by proper choice of design techniques than it is to use corrosion-resistant materials or protective coatings. Sometimes, the combination of both good design techniques and protective coatings is required to achieve the necessary protection for an anticipated corrosion problem.

Design Detail Principles

Figure 4.26[53] demonstrates the following principles to be used in designing automobiles for corrosion resistance:

Principle 1. Keep underbody surfaces dry; avoid ledges, flanges, and pockets where dirt can accumulate and hold moisture. Horizontal flanges should be narrow and should face away from the line of travel. Flanges should slope downward to provide good drainage.

Principle 2. When appearance is of primary importance, use a solder-filled, double-offset lap joint instead of a mastic-sealed coach or lap joint.

Principle 3. Make joints watertight. Use single lap joints to prevent entrance of water. Point lap away from the direction of travel or downward for the best natural drainage.

Principle 4. Seal joints with mastic type compound and cover entire faying surface in riveted joint. Completely seal edges in welded joints to prevent entrance of moisture.

Principle 5. Provide protective flange for lap joint in line with wheel splash to prevent water and road contamination from being driven into area of faying surfaces.

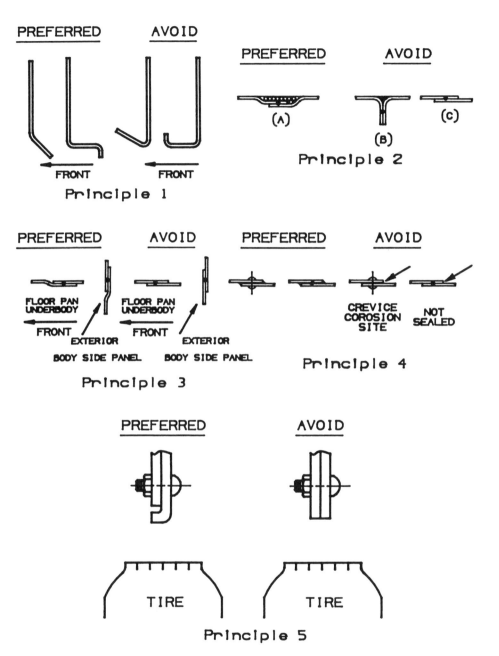

FIGURE 4.26 — *Automotive corrosion design Principles 1 through 11 (figure continued on the next two pages).*

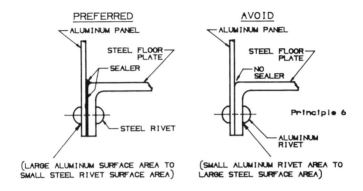

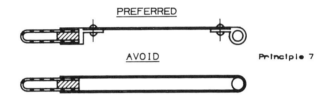

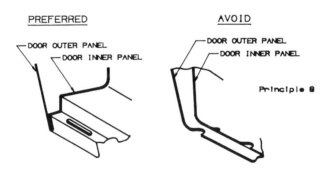

FIGURE 4.26 (continued) — *Automotive corrosion design Principles 1 through 11.*

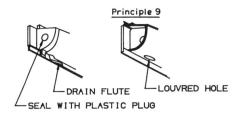

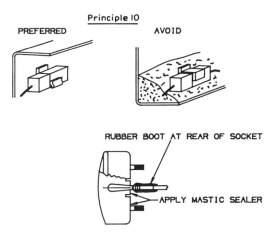

SEALING OF LAMPS TO PREVENT CORROSION

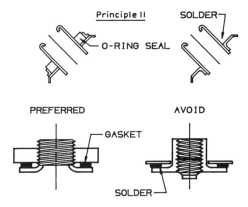

FIGURE 4.26 (continued) — *Automotive design Principles 1 through 11. (Reprinted with permission from the Society of Automotive Engineers, Inc., Warrendale, Pennsylvania, copyright 1982.)*

Principle 6. Avoid using dissimilar metals in contact with each other when possible. To avoid galvanic corrosion, if dissimilar metals must be used, adhere to the following guidelines:

- Use large anode areas (aluminum) to small cathode areas to minimize galvanic effect.
- Use metallic coating (zinc or cadmium) on the anode to reduce galvanic attack.
- Insulate the joint. Use protective coating on both sections, especially on the cathode area.
- Keep joints watertight (use mastic sealer).

Principle 7. Use open construction whenever possible. Avoid box sections and enclosed areas in severe corrosion exposure. Proper application of protective coatings is difficult in confined locations, and it is usually impossible to inspect for completeness of application in enclosed areas.

Principle 8. Provide adequate drainage in doors and in body areas having movable windows. Drain openings should be located to obtain best possible drainage. Openings must be large enough to avoid plugging and to permit drainage flow equal to or greater than entrance flow.

Principle 9. When box sections or enclosed areas are used, provide sufficient openings for applications and drainage of protective coatings. Provision should be made for good ventilation and drainage, such as fluted flanges, with flanges pointing downward and toward the rear to minimize entrance of wheel splash and large enough to avoid plugging.

Principle 10. Keep electrical connectors free from moisture by taking the following steps:

- Place in dry location, preferably in passenger or luggage compartment.
- When mounted externally, use location not exposed to wheel splash and where road salt and dirt will not accumulate.
- Protect externally mounted sockets and connectors with rubber boots.
- Seal joints in lamp housings where moisture may enter. Use mastic type sealer.

Principle 11. Design fuel tank and other fluid containing components to eliminate solder joints that require use of corrosive solder flux.

5 — SPECIFICATIONS AND GUIDES

5.1 Specification Writing for Maximum Corrosion Resistance

Specification Categories

To assure that the designer will actually receive what he orders, he must furnish clear, concise specifications to the manufacturer. If the order is unclear, the supplier may furnish wrong or inadequate material.

Materials of construction for process equipment intended for use in corrosive service are generally specified in the following three broad categories:

1. Chemical composition and mechanical properties;
2. Method of manufacture and heat treatment when required; and
3. Form, dimensional tolerances, and finish.

Regarding the *first category*, chemical composition and mechanical properties, many times the notations CRS (cold-rolled steel) or HRS (hot-rolled steel) have been put on drawings to serve as complete specifications for the steel required. This kind of specification is equivalent to writing down "automobile" on a car order. The buyer may get a Hyundai,[†] and then again, he may get a Cadillac.[†] CRS and HRS could be low, medium, or high carbon steel, alloyed or not. A good example of this occurred when welded towers were ordered and built for an American installation in Mexico. During a storm, the towers fractured and collapsed. The failure was caused by brittle welds formed when medium carbon steel had been furnished for the towers instead of the anticipated low carbon steel. At welded areas, the welding heat had raised the areas around the welds above the lower critical temperature of the steel and when quenching occurred in the air, brittle untempered

---

[†]Registered trade name.

areas were formed that fractured under the stress of the storm. An adequate material specification obviously had not been furnished to the fabricator of the tower.

The *second category*, method of manufacture and heat treatment, is also important. The method of manufacturing, such as welding, brazing, silver-soldering, bolting, riveting, casting, forging, etc., must be specified because this will directly affect the corrosion resistance of the equipment ordered. As discussed before, it is very important that the heat treatment, when required, is carefully specified.

Within the *third category*, dimensional tolerance and finish, the finish is probably the most important from a corrosion standpoint. For example, if a certain finish is required for the corrosion resistance of austenitic or chromium stainless steel equipment, the instructions should be more specific than "a smooth or polished surface is required." Stainless steel suppliers recognize the finish designation Nos. 1, 2D, and 2B for unpolished stainless sheet or plate, and designation Nos. 2BX, 2SF, 3, 4, 6, and 7 for polished stock (in increasing order of polish). One of the finishes listed in Table 5.1 should be specified when a special finish is required.

National Standards

An excellent way for the designer to assure that he will receive the process equipment from the fabricator as it was designed to reduce corrosion is to use national standards. National standards actually represent an agreement between fabricators or suppliers and customers about what can and should be furnished. These standards are not permanent since they require periodic reviews that may result in an amendment, modification, or an otherwise change in a particular standard from year to year. National standards are valuable to the designer because standards:[54]

1. Define what is commercially available together with optional requirements;
2. Provide a convenient reference on company specifications, drawings, and orders;
3. Reduce misunderstandings and minimize disputes; and
4. Represent a production standard that results in a more uniform product, fewer varieties, lower inventories, and lower costs.

There are literally hundreds of standards available for use by the designer. A few of the organizations that publish standards are:

TABLE 5.1 — Finishes for Stainless Steels[1]

Finish No.	Description
	Unpolished
1	Hot rolled, annealed, and descaled
2D	Dull, cold-rolled finish produced by cold rolling, annealing, and descaling
2B	Bright, cold-rolled finish produced by a final light cold roll with polished rolls on stainless sheet with a 2D finish (This is the best finish for subsequent polishing)
	Polished
2BX	No. 2B with a special purpose finish
2SF	Delicate satin finish produced by brushing with a fine abrasive
3	Intermediate polished finish for subsequent finishing operations
4	Following initial grinding with coarser abrasives, final finish with 150 mesh abrasive
6	Dull satin finish produced by Tampico brushing of No. 4 finish sheets with a mixture of oil and abrasive
7	Produced by buffing to a highly polished surface of high reflectivity

[1]From "Fabrication of Stainless Steels" by United States Steel Corp., Pittsburgh, PA 15230.

AA	— Aluminum Association
AISI	— American Iron and Steel Institute
ANSI	— American National Standards Institute
API	— American Petroleum Institute
ASME	— American Society of Mechanical Engineers (Unfired Pressure Vessel Code)
ASTM	— American Society for Testing Materials
AWS	— American Welding Society
AWWA	— American Water Works Association
CDA	— Copper Development Association
CMA	— American Cast Metal Association
MTI	— Materials Technology Institute of the Process Industries
NACE	— National Association of Corrosion Engineers
SAE	— Society of Automotive Engineers
TEMA	— Tubular Exchanger Manufacturers Association

The Federal Government has also developed many standards. For instance, the standards developed by the Department of Commerce acting through the National Bureau of Standards are frequently used by industry, as are standards issued by the Ordinance and Materials Departments of the US Navy, Army, and Air Force. These include standard specifications termed QQS-Federal, MIL-S Army-Navy Aeronautical Specs, and Aerospace Material Specifications (AMS).

Some of these standards and their relevance to the designer are discussed below. For more information, contact the associations directly. (See Appendix for complete addresses of associations.)

ASTM Standards

ASTM standards are probably the most useful to the designer in his endeavor to increase corrosion resistance. ASTM annually publishes standards for a myriad of items from austenitic stainless steel to zinc coatings. The society was formed in 1898 and is a scientific and technical organization formed for the development of standards concerning the performance of materials, products, systems, and services. It is the world's largest source of voluntary consensus standards. As stated in the foreword of the standard, "The Society operates through more than 128 main technical committees. These committees function in prescribed fields under regulations that ensure balanced representation among producers, users, and general interest participants."

Using ASTM standards. Because of this balance of participants on the voluntary committees, any bias is reduced or eliminated. The designer can therefore confidently use these standards knowing that a particular group will not dominate the way standards are written. Using ASTM standards is an excellent way for the designer to assure that he receives what he actually needs.

As an example, suppose the designer wishes to order austenitic stainless steel pipe for a specific plant process requiring materials of construction of excellent corrosion-resisting quality. Since the pipe is to be welded, AISI 304L has been selected as suitable for process use. By examining the annual book of ASTM Standards for 1984, Volume I, under "Steel Pipe," the designer finds 10 standards on austenitic stainless steel pipe. He reviews these standards and determines that ASTM Specification A-312-83 entitled "Seamless and Welded Austenitic Stainless Steel Pipe" best addresses his needs. The specification has 13 sections:

Section I	—	Scope
Section II	—	Applicable Documents (other ASTM and ANSI specifications are referred to)
Section III	—	General Requirements
Section IV	—	Ordering Information (such as quantity, name of materials, process-welded or seamless, grade, size, length, end finish, optional requirements, test report required, ASTM designation, and special requirements or exceptions to the specification)
Section V	—	Materials and Manufacturer (how products are to be manufactured and heat treated)
Section VI	—	Chemical Requirements
Section VII	—	Heat Analysis
Section VIII	—	Product Analysis
Section VIX	—	Tensile Requirements
Section X	—	Mechanical Tests Required
Section XI	—	Lengths
Section XII	—	Finish
Section XIII	—	Marking

At the end of the specification, supplementary requirements, which are not considered unless specified in the order, are listed. These include (S1) Product Analysis; (S2)Transverse Tension Tests; (S3) Flattening Test; (S4) Etching Tests; and (S5) Radiographic Examination.

The designer must understand the standard he expects to use. He cannot just state that the pipe must comply with ASTM Standard A312. A standard is similar to a printed form for a last will and testament; you have to fill in the blanks. The designer's final specifications might read as follows:

> "Furnish 350 ft of austenitic stainless steel to ASTM Standard A312-83 welded AISI 304L, 3-in.-OD (outside diameter) Schedule 10, lengths shall be 15 to 24-ft long, short lengths under 14 ft not accepted, plain ends; test report required. All welds shall be examined radiographically as per S-5."

This kind of specification assures the quality of the piping through manufacture to the final product. The last two numbers on the specifications, such as ASTM Standard A312-83, refers to the date issued (1983). The designer should always refer by date to the specific ASTM standard he specified. A risky notation is "ASTM Standard A312, *latest edition.*" Since the standards are reviewed annually and are often amended or changed, a change may be made which the designer may not need or require.

There are 63 volumes of ASTM standards in 15 sections covering many subjects. Every design office should have the standards that are pertinent to that office's business.

Throughout this book, ASTM standards have been referred to. As mentioned before, they can prove valuable to the designer.

ASTM Corrosion Acceptance Standards

There are also a group of ASTM standards devoted exclusively to acceptance tests for metals. These can be helpful to the designer. Some of these standards are:

1. ASTM Standard A-262-85. Practices for Detecting Susceptibility of Intergranular Attack in Austenitic Stainless Steels (which was discussed earlier).

2. ASTM Standard G-34-79. Test Method for Exfoliation Susceptibility in 2XXX and 7XXX Series Aluminum Alloys.

3. ASTM Standard G-35-73 (1982). Practice for Determining the Susceptibility of Stainless Steels and Related Nickel-Chromium-Iron Alloys to Stress Corrosion Cracking in Polythionic Acids.

4. ASTM Standard G-47-71 (1984). Test Method for Determining Susceptibility to Stress Corrosion Cracking of High-Strength Aluminum Products.

5. ASTM Standard G-48-76 (1980). Test Method for Pitting and Crevice-Corrosion Resistance of Stainless Steels and Related Alloys by the Use of Ferric Chloride Solution.

6. ASTM Standard A-708-71. Recommended Practice for Detection of Susceptibility to Intergranular Corrosion in Severely Sensitized Austenitic Stainless Steel.

7. ASTM Standard G-92-83. Guide for Development and Use of a Galvanic Series for Predicting Galvanic Corrosion Performance.

8. ASTM Standard G-79-83. Guide for Crevice Corrosion Testing of Iron and Nickel Base Stainless Alloys on Seawater and other Chloride-Containing Aqueous Environments.

Any of these standards may be specified by the designer to assure that the material he receives for fabricating process equipment used in critical applications is of the required quality.

The Society of Automotive Engineers (SAE)

The Society of Automotive Engineers (SAE) standards were, of course, originally written for the manufacture of automobiles; however, they are now used extensively in other industries. These specifications are contained in four volumes, namely:

Volume 1 — Materials
Volume 2 — Parts and Components
Volume 3 — Engine, Fuels, etc.
Volume 4 — On-Highway Vehicles and Off-Highway Machinery

Volume 1 is of interest to the designer concerned with corrosion since it incorporates the composition of various materials, metals and alloys, and mechanical properties. In some cases, the response to heat treatment of alloys along with test procedures is referred to. Both ferrous and nonferrous alloys are covered.

The American Society of Mechanical Engineers (ASME) Pressure Vessel Code

The Pressure Vessel Code was written in 1911 by the American Society of Mechanical Engineers (ASME). The purpose was to issue rules for the construction of steam boilers and other pressure vessels. These rules encompassed safety considerations governing the design, fabrication, and inspection of pressure vessels. At the present time, a large code committee exists to write, consider, and act on proposed revisions to this code. In many states, by force of law, pressure vessels must be designed and built in accordance with this code. Such vessels are stamped with the code symbol after passing inspection. The sections in the Boiler Code are:

Section I — Power Boilers
Section II — Material Specifications
Section III — Nuclear Power Plant Components
Section IV — Heating Boilers
Section V — Nondestructive Examination
Section VI — Care and Operation of Heating Boilers
Section VII — Care of Power Boilers
Section VIII — Rules for the Construction of Pressure Vessels
Section IX — Welding and Brazing Qualifications
Section X — Fibre Reinforced Plastic Pressure Vessels
Section XI — Rules for Inservice Inspection of Nuclear Power Plant Components

Probably the most useful sections for the corrosion conscious designer are Sections II, V, VIII, and IX. As with many of the standards, there is an intermingling with other standards. For instance, the specification of base materials given in Section II, Parts A and B of the Code, is identical with those of ASTM. The specification for welding materials given in Section II, Part C, is identical or similar to those of AWS.

The American Petroleum Institute (API)

The American Petroleum Institute (API) Standard 620 ("Recommended Rules for Design and Construction of Large, Welded, Low Pressure Storage Tanks") is used frequently by designers in many industries for low pressure storage tanks for many liquids besides oil. API Standard 650 covers steel storage tanks, specifically for oil storage. The designer can use API standards with confidence when specifying such storage tanks.

American Water Works Association (AWWA)

Many standards of the American Water Works Association (AWWA) are also used throughout industry. The standards for linings on steel, such as AWWA Standard C-203 ("Coal-Tar Protective Coatings and Linings for Steel Water Pipelines") have been particularly useful. Coatings for above ground protection of steel, as embodied in AWWA Standard C-204 ("Chlorinated Rubber — Alkyd Paint System for the Exterior Above Ground Steel Water Piping") have also been very helpful. The AWWA standard for plastic piping, namely Polyvinyl Chloride C-900, Polyethylene C-901, Polybutylene C-902, and Glass-Fibre Reinforced Thermosetting Resin Pressure Pipe C-950, have also been consistently used for specifying plastic pipe for handling many solutions.

The American Welding Society (AWS)

Throughout the years, the American Welding Society (AWS) has consistently upgraded the welding art by issuing recommended practices and procedures for welding various metals and alloys. This has been accomplished through the work of 42 main committees. These committees address fundamentals, inspection and qualifications, certification, processes, safety and health, industrial applications, and several other categories.

The publications produced by the ten committees that cover *industrial applications* are, in this author's opinion, probably of the greatest interest to the designer. The ten committees are:

D1	—	Structural Welding
D3	—	Welding of Marine Construction
D5	—	AWWA-NEWWA-AWS Elevated Steel Water Tanks, Standpipes, and Reservoirs
D8	—	AWS-SAE Automotive Welding
D10	—	Piping and Tubing
D11	—	Welding Iron Castings
D14	—	Machinery and Equipment
D15	—	Railroad Welding
D16	—	Robotic and Automotive Welding

When required, the designer can specify that AWS practices dealing with the welding of specific metals and alloys such as stainless steel, aluminum, and titanium piping (D10C, D10H, D10K) be followed by his suppliers to assure quality welding. The AWS standard addressing the qualification of procedures and operators may also be specified to assure sound welding. (See Chapter 4.1.) The designer can obtain published subcommittee reports or specifications from AWS. (See Appendix for association addresses.) For instance, the *D10 Committee on Piping and Tubing* has nine subcommittees, all of which have issued recommended practices and procedures involving their category of work, except for D10U, which is a new committee. These subcommittees for Piping and Tubing are:

D10C	—	Welding Practices and Procedures for Austenitic Steel Piping
D10H	—	Aluminum Piping
D10I	—	Chromium-Molybdenum Steel Piping
D10K	—	Titanium Piping
D10N	—	Qualification of Procedures and Operators
D10P	—	Local Heat Treating of Pipework
D10S	—	Purging and Root Pass Welding
D10T	—	Low Carbon Steel Pipe
D10U	—	Duplex Stainless Steel Piping and Tubing

The National Association of Corrosion Engineers (NACE)

The National Association of Corrosion Engineers (NACE) was formed in 1943 with the aim of assisting the public and industry in the use of corrosion prevention and control to reduce the billions of dollars lost each year caused by corrosion. This association has many activities all devoted to the field of corrosion. There are over 300 NACE technical committees, each studying various facets of this field. The development of voluntary consensus standards, which encompass recommended practices, test methods, and material requirements is one of the many activities of the NACE technical committees. These should be of interest to the designer in his goal of getting the best quality from corrosion-resistant materials or corrosion control methods and techniques that he may specify. For the complete list of NACE standards, see Reference 55.

The following standards are among those that may be specified by the designer to be followed for his projects concerning corrosion prevention:

1. NACE Standard RP0169-83. "Control of External Corrosion on Underground or Submerged Metallic Piping Systems." Presents procedures and practices which achieve effective control of external corrosion on buried or submerged metallic piping systems. Recommendations are also applicable to many other buried or submerged metallic structures and also describes the use of electrically insulated coatings, electrical isolation, and cathodic protection.

2. NACE Standard RP0275-75. "Application of Organic Coatings to the External Surface of Steel Pipe for Underground Service." Assists users of all types of organic coating materials to obtain a satisfactory application of coating to the external surface of steel pipe.

3. NACE Standard RP0375-75. "Application and Handling of Wax-Type Protective Coatings and Wrapper Systems for Underground Pipelines." Provides information on both hot and cold-applied wax coatings and component wrappers as coatings systems for the protection of underground piping. Suggests proper methods for surface preparation, coating application, handling, storage, and installation of coated piping.

4. NACE Standard RP0178-78. "Design, Fabrication, and Surface Finish of Metal Tanks and Vessels to be Lined for Chemical Immersion Service." Guidelines for design, fabrication, and surface finishing of metal tanks and vessels that are to be lined for corrosion resistance and product contamination prevention.

5. NACE Standard RP0170-85. "Protection of Austenitic Stainless Steel from Polythionic Acid Stress Corrosion Cracking During Shutdown of Refinery Equipment." Examines varying procedures used by industry to protect austenitic stainless steel equipment while idle. Basic protection methods include avoidance of oxygen (air) entry, prevention of liquid water formation, and alkaline wash of surfaces to be exposed.

6. NACE Standard RP0182-82. "Initial Conditioning of Cooling Water Equipment." The procedures outlined are for protecting cooling water equipment from initial corrosion that occurs prior to cooling water exposure and are primarily intended for low carbon (mild) steel equipment, but also apply to other metals and their alloys used in cooling water systems.

7. NACE Standard RP0281-85. "Method of Conducting Coating (Paint) Panel Evaluation Testing in Atmospheric Exposures." A guide for establishing procedures for selecting test panels, preparing and painting the surfaces of panels, selecting test sites, and grading and evaluating panels in atmospheric exposures. These recommendations apply only to atmospheric exposure.

8. NACE Standard TM0274-74. "Dynamic Corrosion Testing of Metals in High Temperature Water." Guidelines for corrosion testing of metals exposed to high temperature water used in high pressure steam plants or water-cooled nuclear reactor plant systems. This standard provides standardized test methods to determine the effects of high temperature water.

9. NACE Standard RP0175-75. "Control of Internal Corrosion in Steel Pipelines and Piping Systems." Describes procedures and practices to achieve effective control of internal corrosion in steel pipe and piping systems in crude oil, refined products, and gas service. This document contains specific practices for protecting steel piping systems.

10. NACE Standard RP0276-76. "Extruded Asphalt Mastic Type Protective Coatings for Underground Pipelines." Guidelines for use of asphalt mastic type coating systems for application to underground pipelines for protection against external corrosion. Material characteristics, application methods, and minimum satisfactory coating requirements are given.

11. NACE Standard TM0174-74. "Laboratory Methods for the Evaluation of Protective Coatings Used as Lining Materials in Immersion Service." Provides guidelines to help manufacturers and users of protective coatings to select materials by providing standard test methods for evaluating protective coatings used as linings for immersion service. This standard provides two test methods for evaluating protective coating on any substrate, such as steel, copper, aluminum, etc., so both the factors of chemical resistance and permeability can be considered.

12. NACE Standard TM0169-76. "Laboratory Corrosion Testing of Metals for the Process Industries." Guideline for laboratory immersion corrosion studies. Provides a consensus on the best current technology for laboratory corrosion testing.

13. NACE Standard RP0272-72. "Direct Calculation of Economic

Appraisals of Corrosion Measures." Establishes standard methods for economic appraisal of alternate corrosion control measures. Use of the calculation schedules in the standard should permit the corrosion engineer to prepare an acceptable corrosion control proposal (discussed in Chapter 1.2).

14. NACE Standard TM0270-70. "Method of Conducting Controlled Velocity Laboratory Corrosion Tests." Provides a method by which laboratory corrosion tests may be reproducibly conducted under velocity conditions. The method described is applicable only to the study of the effects of solution velocity on corrosion rate of a metal and is not to be used to determine the effect of a solution impinging on a metal surface (discussed in Chapter 2.2).

Company Standards

Because of repetitive demand for certain items not completely covered by national standards, companies have produced their own standards. There also may be certain situations companies have to deal with that are not covered at all by national standards. A special piping system for handling sulfuric acid is a good example of this type of standard. In small companies, national standards are sometimes modified to satisfy this need. However, standards sections are sometimes available in large companies to help the designer. In one large chemical company, one group administers company standards. Thirty-one subcommittees, covering many areas such as welding, insulation, plastics, heat exchangers, and protective coatings, write the specifications and review and update them periodically. Since these specifications address the unique problems of a company, they can of course be beneficial to the company designer.

What the Designer Should Remember When Writing Specifications

When writing specifications, the designer should remember:

1. Make specifications as short as possible, but they must clearly define what is required and at what quality level. The quality level provides assurance that process equipment will perform reliably and will not fail by corrosion prematurely.

2. Avoid vague statements such as "all equipment and piping after welding must be stress relieved." For instance, when field connections are to be made, stress relieving after welding is difficult, so flanged connections may be substituted that require

no welding. If the blanket statement above should not have included the field connections, cost would have been raised unnecessarily.

3. Do not simply specify that x-ray quality welds are required without defining the level of quality required for acceptance.

   Example — Miles of 3-in. (76-mm) diameter welded AISI 304L austenitic stainless steel pipes were produced by a single manufacturer. The original purchase specification required 100% radiographing of all the longitudinal weld seams. When lengths of this pipe were field welded into fittings, the field welds were radiographed, which revealed not only the field welds themselves, but short portions of the longitudinal welds, which, in many cases, were very poor. The pipe was cut out, and the pipe fabricator was contacted. The fabricator stoutly maintained that all of his pipe was x-ray quality and he pointed to a cabinet full of radiographs. We read these radiographs and found a lot of evidence calling for many rejections and repairs. It was finally disclosed that the pipe fabricator had not viewed any of the radiographs. He thought that passing the x-rays through the welds made them x-ray quality! This story is not a fabrication! Luckily, only a small amount of pipe had actually been field welded and all the poor welds were identified and the pipe was rejected.

   The designer must not only define the radiographic indications that will be cause for rejection, but must specify the extent of random inspection such as 5, 10, 33, or 100% of the pipe welded. Radiographic inspection is discussed in Chapter 6.1.

4. Consider costs when writing specifications. The specification should not be so restrictive that satisfactory quality material will be excluded. In addition, specifications should not restrict the manufacturers so much that his costs, and hence the price, will be unnecessarily high. On the other hand, the specification must not be so vague that inferior quality may be allowed.

5. Make safety paramount in any specification. For instance, if pneumatic testing must be conducted (avoid such testing, if possible), all welds (that can be) should be radiographed before testing. All welds that cannot be radiographed because of geometry should be tested with the liquid penetrant or magnetic

particle method also before testing. (If a break occurs during pneumatic testing, shattering of the equipment tested can occur.)

6. Whenever possible, to hold down costs, have the equipment made with commercially available materials using standard methods of construction. As an example of when this was not done, some years ago, a bellow-sealed valve was required for an extremely hazardous service. The company involved, now defunct, spent a large amount of the taxpayer's money to design and build a bellow-sealed valve when a satisfactory and better valve had been available all along on the shelf from a major valve manufacturer at a much lower price.
7. Before finalizing a specification, have the specification reviewed by potential fabricators. Many times, a fabricator can suggest ways to save money because he certainly knows best how to build his product. Specify primarily how the equipment should perform rather than detailing how the fabricator should build it.
8. Specifically note the tests required to assure quality.
9. Carefully note in the specification the equipment that is to be inspected during its manufacture so that the fabricator can make provisions for the inspector's visits at the proper time.
10. Do not hesitate to specify a trade name or a catalog number for a product if that product will do the required job. This, of course, precludes that the manufacturer is reliable, competent, and has the reputation of fabricating products of consistent quality at competitive cost.

☐ 5.2 Stainless Steels

Stainless steels are a very important class of alloys that are extensively used in industry. A noncorroding, nonrusting steel was sought for many years. The "discovery" of stainless steel during the period 1900 to 1915 was preceded by more than a century of cumulative effort by many men. The remarkable fact is not that it was discovered in the early 1900s, but that is was not discovered many years before.

In 1790, Vauguelin discovered chromium by reducing it from its ore. Chromium was added experimentally to steels during the 1800s, but by some strange quirk of fate, too little chromium was added to make steels that were corrosion resistant. As a matter of fact, in 1892, Hadfield concluded that chromium *impaired* the corrosion resistance of steel. His alloy contained a maximum of 9% chromium, which is just below the minimum amount required for corrosion resistance. In 1913, Brearly was investigating new alloys for the lining of naval guns. He noticed, during a metallographic study, that the customary etching reagents did not etch steels having high chromium contents. He patented a type of stainless steel in 1916 that is virtually the present day AISI 420 alloy. Almost at the same time, 1909 to 1912, Maurer developed an *austenitic* stainless steel, 18%Cr-8%Ni. From then on, development of stainless alloys was practically continuous, though, at some times, rather hectic. Around 1920, One of history's greatest lawsuits was filed. It was the suit of American Stainless Steel Co. and Electro Metallurgical Co. vs Rustless Iron Corp. of America over who discovered 17% chromium alloys.

Role of Chromium

Chromium is the resistant keystone of all the types of stainless steel. When the amount of chromium is gradually increased in steel, a drastic change in corrosion resistance results. As the chromium content is increased under oxidizing conditions, a surface film attempts to form and then it rather suddenly completes itself when about 12% chromium is present. The oxide layer at that concentration of chromium is tenacious and impenetrable to migrating atoms, either oxygen atoms pressing inward or metallic atoms pressing outward, thus passivating the corrosion action. (Conversely, when the film is destroyed by certain chemical or mechanical action, the steel activates and becomes liable to attack, as discussed previously.) When this strong, continuous film has been formed, the metal is in the *passive* state.

Types of Stainless Steels

Since there are four types of stainless steel to select from, it is important for the designer to understand the differences in corrosion resistance between them. The four general types are austenitic, martensitic, ferritic, and precipitation hardening. Refer to Tables 5.2 through 5.5 for classes, type numbers, and various uses for these types of stainless steels. Comments on the corrosion resistance of various metals and alloys that follow in this section are to highlight certain individual characteristics in widely used services. However, in order to determine the corrosion resistance of various types of stainless steels to contemplated process solutions, the designer should have corrosion tests made following procedures outlined in Chapter 2.

The four classes of stainless steel vary considerably in their chemical composition, corrosion resistance, mechanical properties, and

TABLE 5.2 — Austenitic Stainless Steels

Class	Hardenability	AISI Type	Analysis Compared to Basic Type	Various Uses (Applies to Listed S.S. Types)
Chromium-Nickel	Hardened Only by Cold Work	301	Chromium and nickel lowered for more work hardening	**Transportation Industry:** Trim, bumpers, wheel, covers for autos, sides for railroad cars.
		302	Basic type 18% chromium, 8% nickel	**Food Industry:** Tanks, bottles, pasteurizers, buckets, bottling equipment, and vats.
		304	Carbon lower than the basic type	**Laundries:** Kettles, tables, and driers.
		304L	Carbon below 0.03% to avoid carbide precipitation caused by welding	**Dairies:** Pasteurizers, separators, tanks, and coolers.
		321 347	These two alloys have been in general replaced by AISI 304L; these alloys are stabilized with titanium (321) and niobium (347) to avoid carbide precipitation caused by welding	**Pharmaceutical:** Tanks, kettles, and processing equipment. **Ships:** Hatch covers, doors, and stacks. **Building Industry:** Architectural trim and paneling. Excellent resistance to the atmosphere. Kitchen sinks, trim, etc. **Chemical Industry:** Heat exchangers, condensers, tanks, sieve plates, pumps, low-temperature use, dye and bleach equipment for textiles, piping, and tubing. Tanks, tubing, etc., for handling liquid nitrogen and oxygen. Used for storing nitric acid. (Not AISI 301 or 302.)
		303	Sulfur or selenium added for easier machining	For producing nuts, bolts, screws, and shafts where a free-machining steel is required; in certain environments, these materials have limited corrosion resistance.
Chromium-Nickel	Hardened Only by Cold Work	305	Nickel higher for less work hardening	Drawing, cold heating, and spinning applications.
		308	Nickel and chromium higher	Mainly for electrodes or filler metal to weld AISI 304 SS to compensate for any losses of chromium and nickel during welding.
		308L	AISI 308 carbon content reduced	Welding AISI 304L where carbon must be below 0.030%.
		309 310	Chromium and nickel still higher for more corrosion and scaling resistance	Same uses as noted for AISI 304, 304L, etc., for chemical industry, including processing equipment; in addition, these alloys resist oxidation up to 1900 to 2000 F, such as annealing boxes, resistors, retorts, and furnace parts.
		309Cb	Niobium and tantalum added to avoid carbide precipitation	Welding electrodes and filler metal piping and tubing.

TABLE 5.2 — Austenitic Stainless Steels (continued)

Class	Hardenability	AISI Type	Analysis Compared to Basic Type	Various Uses
Chromium-Nickel	Hardened Only by Cold Work	316 317	Molybdenum added for more resistance to pitting and corrosion	**Chemical Industry:** These types are used where AISI 304 may not have adequate corrosion resistance; can be used for most of the uses listed for AISI 304, 304L, etc., for chemical industry, including towers and screens. Has better resistance to hot organic acid; however, resistance to nitric acid is substantially less. Better resistance to dilute sulfuric acid than AISI 304. Used for handling a wide range of corrodents in piping, tanks, heat exchangers, pumps, etc. **Paper Industry:** Pumps, evaporators, digesters, piping, and tubing. **Soap Manufacturing:** Piping and tubing, tanks, towers, heat exchangers.
Chromium-Nickel	Hardened Only by Cold Work	316L	Carbon content limited to 0.30%	Same applications as listed for AISI 316, where protection from carbide precipitation from welding is required.
Chromium-Nickel Manganese	Hardened Only by Cold Work	201[1]	Chromium and nickel lower for more work-hardening	Same uses as for AISI 301.
Chromium-Nickel	Hardened Only by Cold Work	202[1]	Chromium 18% Nickel 5% Manganese 8% basic type	Same uses as for AISI 302.
		204[1]	Carbon somewhat lower	Same uses as for AISI 304.
		204[1]	Carbon below 0.03% to avoid carbon precipitation caused by welding	Same uses as for AISI 304L (can be welded with AISI 308L electrode or filler metal.)

[1]During a period of nickel shortages, these types were developed to produce basically as much austenite by substituting manganese. They have about the same corrosion resistance as their counterparts in the 300 Series, but have higher strengths.

TABLE 5.3 — Martensitic Stainless Steels

Class	Hardenability	AISI Type	Analysis Compared to Basic Type	Various Uses
Chromium-Iron	Hardenable by Heat Treatment	403	12% chromium with higher mechanical properties	Turbine valves and turbine buckets.
		410	Basic type 12% Chromium	Furnace parts and other heat-resisting uses to 1400 F; trim and seats for steel and cast iron industrial valves; shafts and impellors for pumps; tubing, bubble caps, and petroleum towers; coal screens; bushings and bearings, nuts, and bolts; cutlery, tableware, and kitchen tools.
		414	Nickel added to increase corrosion resistance and and strength	
		416	Sulfur or selenium added for free machinability	Screws, valve stems, nuts, and bolts are made from this metal to save on machining costs. Has reduced corrosion resistance because of the presence of sulfur or selenium.
		420	Carbon higher for greater hardenability	Heat-treated springs and cutlery of all types; heat-treated bushings, bearings, valve seats, and trim; surgical instruments. This alloy is the first SS and was invented by Brearly in 1913.
		431	Higher chromium and nickel for increased corrosion and mechanical properties	Aircraft parts; shafts for ships; milk separators; valves in milk lines; pump parts; wires requiring stiffness and strength.
		440 A, B, and C	Carbon higher to give great hardness	Parts requiring hardness up to 65 Rockwell C; cutlery, valve seats, and trim; heat-treated springs; bushings and bearings; gages of various kinds and surgical instruments.

TABLE 5.4 — Ferritic Stainless Steels

Class	Hardenability	AISI Type	Analysis Compared to Basic Type	Various Uses
Chromium-Iron	Non-Hardenable	405	Aluminum added to 12% Chromium to prevent hardening	Boiler tubing and heat exchanger tubing that are welded but not heat treated afterwards.
		430	Chromium 17% — basic type	Heat-resistant parts subject to low stress at temperatures to 1550 F, including furnace and retort parts; heat exchangers, condensers, bubble caps, piping, and tubing; tanks for transportation and storage; automotive trim, and molding.
		431F	Sulfur or selenium added for machinability	Nuts, bolts, screws, and other parts where easy machining is required; has lowered corrosion resistance
		442	Chromium higher to increase scaling resistance	Various equipment to resist temperatures up to 1750 F at <u>low stress</u>, such as furnace parts, soot-blower elements, and heat recuperators; good resistance to high-temperature oxidation and sulfur attack.
		446	Chromium much higher for improved scaling resistance	Stack dampers, burner nozzles, kiln lining, other furnace parts including baffles; also has good resistance to high-temperature oxidation and sulfur attack.

TABLE 5.5 — Precipitation-Hardening Stainless Steels

Class	Hardenability	AISI Type	Analysis Compared to Basic Type	Various Uses
Precipitation Hardening	Hardenable by Heat Treatment	17-4 PH	16% Cr4Ni plus Cu and Ta	High-strength materials, up to around 200,000 psi strengths in the treated condition; useful for missile and aircraft industries where high strength and resistance to mildly corrosion condition is required.
		17-7 PH	A little higher Cr with about double the nickel content plus 1% Al	
		pH 15-7 Mo	Lower Cr with high nickel 7% plus 3.5% Mo at 1% Al	Good resistance to atmospheric conditions.
		Stainless W	17 Cr 7 Ni plus Ti and Al	Used for rubbing parts because of its high hardness including valve seats, valve discs, bushings, and bearings.

hence, uses. Nickel, manganese, carbon, and nitrogen are austenite formers while chromium, molybdenum, and silicon are ferrite formers. The structure of these stainless steels will depend on their chemical composition.

Austenitic stainless steels. The 200 and 300 Series contain nickel as the principal *austenite* former. During World War II, the 200 Series stainless steels were developed because of the shortages of nickel, with manganese substituting in part for the nickel. Both series have about the same corrosion resistance (200 and 300 Series, Table 5.2). The austenitic stainless steels are more corrosion resistant than the straight chromium grades of stainless steel and generally have the best corrosion resistance of the four types of stainless steels. The 200 and 300 Series stainless steels are hardenable only by cold working. The bulk of the stainless steels used by the process industries are AISI 304, 304L, and 316. The 300 Series stainless steels are probably the most used stainless steels.

Martensitic stainless steels. Martensitic stainless steels (Table 5.3) contain chromium that acts as a *ferrite* former. It is hardenable by heat treatment, the same as carbon steel, because it can form martensite, which is a hard constituent. The corrosion resistance tends to be better in the hardened condition than the annealed condition. The corrosion resistance is generally not as good as austenitic or ferritic stainless steels. The martensitic steels AISI 420, 431, 440A, 440B, and 440C are used when a high hardness coupled with corrosion resistance is required.

Ferritic stainless steels. The ferritic stainless steels (Table 5.4) cannot be hardened by heat treatment because they do not undergo a phase transformation at higher temperatures. These steels are excellent for high temperature service; however, they do not show consistent structural strength at elevated temperatures, so stresses must be kept low. AISI 442 and 446 are generally used in heat treating equipment. AISI 430 was the original material of construction used by the chemical industry for nitric acid service; however, austenitic stainless steels have largely supplanted AISI 430 for this service because of their better fabricability and lower corrosion rates. Ferritic stainless steel alloys have a substantial advantage over austenitic stainless steels because they generally handle stress corrosion cracking better, particularly when chlorides are present.

Precipitation-hardening stainless steel. Precipitation-hardening stainless steel (Table 5.5) hardens by solution-quenching followed by aging, which places precipitates in the slip planes of the alloys rendering these metals hard. None of the precipitation-hardening alloys listed in Table 5.5 have outstanding corrosion resistance. Their corrosion resistance, for instance, is less that of AISI 304 stainless steel. When a very strong, hard alloy is required and only mild corrosion resistance is necessary, these alloys may be ideal. Tensile strengths of 200,000 psi may be obtained. Table 5.6 denotes the chemical composition of wrought stainless alloys, including those discussed above.

Cast stainless alloys. Cast stainless steels do not fall conveniently into any one of the stainless steel types because they vary somewhat from the chemical compositions of wrought alloys, as described before.

TABLE 5.6 — Chemical Composition of Wrought Stainless Steels

AISI Type	C	Mn Maximum	P Maximum	S Maximum	Si Maximum	Cr	Ni	Mo	Other Elements
201	0.15	6.5	0.045	0.030	1.00	17	4.5		N 0.25 maximum
202	0.15	8.0	0.045	0.030	1.00	18	5.0		N 0.25 maximum
204	0.08	8.0	0.045	0.030	1.00	18	5.0		N 0.25 maximum
204L	0.03	8.0	0.045	0.030	1.0	18	5.0		N 0.25 maximum
301	0.15 maximum	2.00	0.045	0.030	1.00	16.00/18.00	6.00/8.00		
302	0.15 maximum	2.00	0.045	0.030	1.00	17.00/19.00	8.00/10.00		
303	0.15 maximum	2.00	0.20	0.15 minimum	1.00	17.00/19.00	8.00/10.00	0.60 maximum	
304	0.08 maximum	2.00	0.045	0.030	1.00	18.00/20.00	8.00/12.00		
304L	0.03 maximum	2.00	0.045	0.030	1.00	18.00/20.00	8.00/12.00		
308	0.08 maximum	2.00	0.045	0.030	1.00	19.00/21.00	10.00/12.00		
309	0.20 maximum	2.00	0.045	0.030	1.00	22.00/24.00	12.00/15.00		
310	0.25 maximum	2.00	0.045	0.030	1.50	24.00/26.00	19.00/22.00		
316	0.08 maximum	2.00	0.045	0.030	1.00	16.00/18.00	10.00/14.00	2.00/3.00	
316L	0.03 maximum	2.00	0.045	0.030	1.00	16.00/18.00	10.00/14.00	2.00/3.00	
317	0.08 maximum	2.00	0.045	0.030	1.00	18.00/20.00	11.00/15.00	3.00/4.00	
321	0.08 maximum	2.00	0.045	0.030	1.00	17.00/19.00	9.00/12.00		Ti 5xC minimum
347	0.08 maximum	2.00	0.045	0.030	1.00	17.00/19.00	9.00/13.00		Cb-Ta 10xC minimum
403	0.15 maximum	1.00	0.040	0.030	0.50	11.50/13.00			
410	0.15 maximum	1.00	0.040	0.030	1.00	11.50/13.00			
405	0.08 maximum	1.00	0.030	0.030	1.00	11.50/14.00			Al 0.10/0.30
414	0.15 maximum	1.00	0.040	0.030	1.00	11.50/13.50	1.25/2.50		
416	0.15 maximum	1.25	0.06	0.15/0.35	1.00	12.00/13.50			
420	over 0.15	1.00	0.040	0.030	1.00	12.00/14.00			
430	0.12 maximum	1.00	0.040	0.030	1.00	14.00/18.00			
431	0.20 maximum	1.00	0.040	0.030	1.00	15.00/17.00	1.25/2.50		
440A	0.60/0.75	1.00	0.040	0.030	1.00	16.00/18.00		0.75 maximum	
440B	0.75/0.95	1.00	0.040	0.030	1.00	16.00/18.00		0.75 maximum	
440C	0.95/1.20	1.00	0.040	0.030	1.00	16.00/18.00		0.75 maximum	
446	0.20 maximum	1.50	0.040	0.030	1.00	23.00/27.00			N 0.25 maximum
17-4 pH						16.5	4		4 Cu; 3Cb + Ta
17-7 pH						17	7		1.1 Al
pH 15-7 Mo						15	7		2.5 Mo, 1.2 Al
Stainless W	0.07					17	7		0.7 Ti, 0.2 Al

The reason for this is that to be castable, an alloy must have a higher silicon content. (Wrought alloys are usually limited to 1% Si.) The additional silicon content is required particularly for thin cast sections. Also, high ferrite content is acceptable in cast products because rolling and forming is not required. Because the ferrite content in an austenite matrix substantially increases strength, such mixed structures of ferrite and austenite would cause great difficulty in rolling and forming if an attempt was made to treat these alloys as wrought alloys, therefore limiting this type of stainless steel to the cast form.

The amount of the ferrite phase in cast alloys is increased by adding or increasing the amounts of ferrite formers, such as chromium and molybdenum, and/or by decreasing the amounts of austenite formers and stabilizers such as nickel, nitrogen, and carbon. Because the cast stainless steel alloys are not limited to chemical composition like wrought alloys, they have unusual properties such as increased yield strengths. Since these alloys can have two distinct microstructures, they are called *duplex* alloys.

Table 5.7 shows some of these cast stainless steels. Note that a designation starting with a "C" indicates that the alloy is used for

TABLE 5.7 — Cast Stainless Steels

Cast Alloy Designation	Wrought Alloy Comparison	% Composition (bal Fe)		
		Cr	Ni	Other Elements
CA-15[1]	410	11.5 to 14	1 maximum	
CA-40[1]	420	11.5 to 14	1 maximum	
CC-50[1]	446	26 to 36	4 maximum	
CF-8[1]	304	18 to 21	8 to 11	
CF-8[1] M	316	18 to 21	9 to 12	Mo 2.2 to 3
CF-20[1]	302	18 to 21	8 to 11	
CF-8[1] C	347	18 to 21	9 to 12	Co (8 × C) minimum (1.0) maximum
CF-16[1] F	303	18 to 21	9 to 12	Mo 1.5 maximum Se 0.20 to 0.35
CD-4 M Cu	—	25 to 27	4.75 to 6	2 Mo, 3 Cu
CH-20[1]	309	22 to 26	12 to 15	
CK-20[1]	310	23 to 27	19 to 22	
CN-7[1] M	Alloy 20	18 to 22	21 to 31	varying amounts of Si, Mo, and copper depending on producer
HC	446	26 to 30	4 maximum	0.50 C
HF	302	19 to 23	9 to 12	0.20 to 0.40 C
HH	309	24 to 28	11 to 14	0.20 to 0.50 C 0.20 maximum N
HK	310	24 to 28	14 to 18	0.20 to 0.60 C

[1]Number refers to "points" of carbon content. For instance, 7 is 0.07%; 20 is 0.20%.

aqueous services and an "H" for heat-resistant services.

The corrosion resistances of cast and wrought stainless steels of similar composition are considered equivalent in most environments. Table 5.7 shows the wrought alloy composition type number it most closely resembles for comparison. "C" alloys contain a number in their designation that signifies the number of points of carbon in the alloy. A point of carbon is 0.01%.

"C" alloys are used for most of the uses noted for wrought alloys. "H" alloys have widespread use in heat-resistant applications, such as castings for turbo superchargers, gas turbines, power plant equipment, and equipment used for the manufacture of glass, cement, synthetic rubber, petroleum products, and chemicals.

Corrosion Resistance of Stainless Steels

Nitric acid. The 300 Series stainless steels are extensively used for handling nitric acid. The most economical choice is generally AISI 304L, which shows good corrosion resistance in the as-welded condition up to about 55 to 60% at 220 F (104 C). The storage of nitric acid at ambient temperatures is handled by AISI 304L up to about 92% acid. (Above that concentration, either AA 3003 or 1100 aluminum has a lower corrosion rate and is generally used.)

AISI 430 stainless steel has been used extensively in the past for shipping and handling nitric acid but has been largely replaced by AISI 304L stainless steel because its resistance covers a broader range of concentrations. AISI 430 is occasionally used when austenitic stainless steels are not satisfactory because of stress corrosion cracking resulting from the presence of chlorides. The major drawback with AISI 430 stainless steels is that they must be heat treated after fabrication. In addition, they are somewhat difficult to weld and must be heat treated after welding or extensive cracking will result. AISI 446 is as resistant to nitric acid as austenitic stainless steels and has the advantage of not being susceptible to chloride stress corrosion cracking. However, it has a great drawback because, for all practical purposes, it is unweldable. Its use for this reason is confined to seamless tubing and pipe not requiring welding.

AISI 316 has been found to be less resistant to nitric acid than other members of the 300 Series. AISI 309 and 310 are not greatly superior in corrosion resistance to AISI 304 in boiling 65% nitric acid. However, a special low carbon version of AISI 310 is extensively used for hot nitric acid.

Hydrochloric acid. Hydrochloric acid rapidly corrodes all four types of stainless steels.

Phosphoric acid. AISI 316L can be used in up to 70 to 80% phosphoric acid if the temperature is below 220 F (104 C). AISI 304L can be used for dilute solutions at ambient temperatures. However, when commercial solutions contain fluoride impurities, the corrosion resistance of austenitic stainless steels will be much lower.

Organic acids. AISI 316L can be used for most organic acids at all concentrations up to the boiling point. Oxalic and formic acids are the exceptions. Oxalic can only be tolerated by AISI 316L at ambient temperatures while formic may be tolerated only up to around 200 F (93 C). At temperatures above the atmospheric boiling point, a more resistant alloy, such as Hastelloy[†] C-276, should be selected.

Sulfuric acid. As can be seen by consulting the Nelson charts, the 300 Series stainless steels can be used when sulfuric acid is either very dilute or is very strong. In dilute sulfuric acid, the 300 Series stainless steel show borderline passivity where a small change in temperature or concentration can cause very severe corrosion. When reducing agents are present, the protective film may be destroyed causing increased corrosion. Offsetting this, if oxidizing agents are present, the film will be preserved and corrosion resistance will be assured. AISI 316L is much more resistant than AISI 304L to sulfuric acid. AISI 304L is only resistant to cold acid at dilutions below 1%. AISI 316L is satisfactory in cold acid to about 10%, as long as the acid is aerated to preserve the passive state.

☐ 5.3 Nickel and Nickel Alloys

Table 5.8 shows the normal composition of nickel and nickel alloys and some uses for these versatile materials. Each of these alloys have excellent corrosion resistance varying in corrosion-resistant properties, as noted below. All of these alloys are more expensive than 300 Series stainless steels and therefore, should not be specified by the designer when stainless steel is suitable.

Corrosion Resistance of Nickel 200

Nickel is a very versatile corrosion-resistant metal. It combines the mechanical properties of the order of mild steel and relatively high corrosion resistance with favorable welding and fabrication qualities. Other characteristics of nickel are that it is:

[†] Registered trade name.

TABLE 5.8 — Nominal Composition and Uses of Nickel and Nickel Alloys

Alloy Name	Percent Nominal/Composition						Various Uses
	C	Cr	Ni	Mo	Cu	Fe	Other	
200	0.08		99+					Caustic evaporators and caustic fusion pots; for a variety of chlorination type reactions and for handling chlorine, hydrogen chloride, chlorinated hydrocarbons; heating coils, tanks, tubular condensers, and evaporators.
201	0.01		99+					For use in caustic and other applications involving temperatures above 600 F; the low carbon avoids harmful graphite precipitation.
Monel 400	0.12		67		32	1		Evaporators, piping, reaction vessels; paper mill and oil refinery equipment; storage tanks, food handling equipment, nozzles, bushings, and turbine blading.
Inconel 600	0.08	15	76			8		Process equipment, dairy equipment, manifolds, and food handling equipment; used extensively in oxidizing and carburizing atmospheres at high temperatures.
Inconel 625	0.02	21	61	9		3	4 Co	For process equipment requiring high-strength and high-temperature corrosion resistance.
Incoloy 800	0.05	21	33			46	0.3 Ti / 0.3 Al	Pigtails for reformer furnaces and other high-temperature corrosion applications.
Carpenter 20-Cb-3	0.04		34	2.5	3.5	38	0.8 Cb	Heat exchangers and heat exchanger tubes, mixing tanks, process piping, bubble caps, metal cleaning and pickling tanks; pumps, valves, and fittings.
Incoloy 825	0.03	21	42	3	2.3	30	0.9 Ti	Mixing tanks, process piping, valves, and fittings.
Hastelloy B-2	0.01	1	bal	28		2	Co 1 / Si 0.1 / Mn 1	Process equipment, piping, tubing. Reaction vessels, and heat exchangers.
Hastelloy C-276	0.01	15	bal	16		5.5	Mn 1 / Si 0.08 / W 4	Tanks, heat exchangers, piping, tubing, and process equipment.
Hastelloy G-3	0.015	22.2	bal	7		19.5	Co 5 / W 1.5 / Si 0.4 / Mn 0.8 / Cu 1.8 / Cb plus Ta 0.3	Pulp digesters, dissolver vessels and piping, tanks.
Hastelloy G-30	0.03	30	bal	5.0	2	15	Co 5 / W 3 / Si 8 / Mn 1.5 / Cb plus Ta 1	Resistant to phosphoric acid and highly oxidizing acids such as nitric/hydrochloric, etc. Useful in pickling acid operations and gold extraction.
Hastelloy C-22	0.01	22	bal	13		3	Co 2.5 / W 3 / Si 0.08 / Mn 0.5 / V 0.35	Highly pit resistant. Excellent resistance to oxidizing aqueous media. Resistant to strong oxidizers such as ferric and cupric chlorides.

1. Resistant to neutral or alkaline salts unless they are strongly oxidizing, such as with sodium hypochlorite.
2. With the exception of silver, it has the best resistance to alkalines and is widely used for caustic evaporators and caustic fusion pots.

3. Resistant to anhydrous ammonia but is attacked by ammonium hydroxide solutions.
4. Not susceptible to stress corrosion cracking but, in some applications, is susceptible to pitting under deposits in brackish water.
5. Widely used for many chlorination-type reactions and for handling chlorine, hydrogen chloride, and chlorinated hydrocarbons.
6. Attacked by sodium hypochlorite.
7. Widely used in the food industry.
8. Attacked and embrittled by sulfur-bearing gases at elevated temperatures, as are its alloys. (See Figure 3.77 previously referred to.) Susceptible to both general and intergranular corrosion in sulfur-bearing gases at temperatures above 600 F (315 C).
9. Resistant to steam but may be severely attacked by condensates.
10. Only moderately resistant to some acids but is badly attacked by oxidizing acids, especially nitric acid.
11. Resistant to neutral and slightly acid solutions.

Nickel 201

For caustic or other services involving temperatures above 600 F (315 C), the low carbon grade Nickel 201 should be specified instead of Nickel 200 to avoid harmful graphite precipitation. However, the presence of sulfur in process solutions will cause serious attack on both Nickel 200 and 201.

Alloy 400

Characteristic of Alloy 400 (Monel[†] 400) is that it is:

1. Not very different in corrosion resistance from nickel, but since it is less expensive and appreciably stronger, it is advantageous for the designer to use it instead of nickel when either alloy would be satisfactory.
2. More corrosion resistant than nickel under reducing conditions. In certain environments where reducing compounds are present, or which are nitrogen-blanketed, even strongly acidic solutions can be handled satisfactorily by this alloy.

[†]Registered trade name.

3. Offers some resistance to hydrochloric and sulfuric acids under modest conditions but is seldom used.
4. Used for caustic solutions, because it is not susceptible to SCC.
5. Chloride salts do not cause SCC.
6. Widely used for containing hydrofluoric acid solutions. However, corrosion rates are substantially increased in the presence of air. When used for hydrofluoric acid, this alloy must be stress relieved to obviate stress corrosion cracking.

Alloy 600

Characteristics of Alloy 600 (Inconel[†] 600) are that it is:

1. Very resistant to stress corrosion cracking.
2. Resistant to mixtures of steam, CO_2 and air. It does not experience the limitations as does nickel or Monel in this service.
3. Not attacked by ammonium hydroxide solutions as are nickel and Alloy 400.
4. Almost as resistant to alkalis as nickel and is somewhat less vulnerable to the presence of sulfur compounds or sulfur-bearing gases; however, Alloy 600 welded equipment should be stress relieved if it is to be used in strong, hot caustic environments.
5. Only fairly resistant to mineral acids but is used extensively for handling fatty acids at high temperatures.
6. Stronger than nickel or Alloy 400 and is more resistant to oxidation. It is also resistant to chlorine and chlorine-containing compounds at high temperatures. In some cases, it is preferable over nickel because of its higher strength and oxidation resistance.

Alloy 625

Characteristics of Alloy 625 are that it is:

1. Extremely resistant to oxidizing and carburizing atmospheres at high temperatures.
2. Highly corrosion resistant for applications involving halogen compounds.
3. Very corrosion resistant to stress corrosion cracking.

[†]Registered trade name.

Alloy 800

Alloy 800 (Incoloy[†] 800) differs from AISI 310 stainless steel (25% Chromium, 20% Nickel) primarily because of its higher nickel content (33%).

1. The added nickel improves resistance to chloride stress cracking.
2. It reduces the tendency for sigma phase formation. (Sigma phase can cause embrittlement in high-temperature service.)
3. Alloy 800 is susceptible to intergranular attack if used in the as-welded condition in severely corrosive processes.

Carpenter 20 Cb-3

Characteristics of Carpenter 20 Cb-3 are that it is:

1. Similar in nickel content to Alloy 800 but also contains molybdenum, copper, and columbium so it is resistant to a wider range of process conditions. Because it is stabilized, it can be used in the welded condition.
2. Superior to the 300 Series stainless steels in resistance to sulfuric acid and can be used over a considerable range of concentrations and temperatures.
3. Used successfully in many severely corrosive applications. The most frequent services are in hot processes containing sulfuric acid.
4. Used widely for applications where a 300 Series stainless steel would be satisfactory on the process side but would fail by chloride stress corrosion cracking on the cooling water side.
5. Stabilized with columbium.

(Carpenter 20 Cb-3 is sometimes classified as a stainless steel.)

Alloy 825

Characteristics of Alloy 825 are that:

1. Its corrosion resistance is generally about the same as that of Carpenter 20 Cb-3.
2. It is stabilized with titanium and is consequently not susceptible to intergranular attack.
3. It contains more nickel than Carpenter 20 Cb-3. Under certain conditions, therefore, it will show increased corrosion resistance. However, there are other conditions where Carpenter 20-Cb-3 is

---

[†]Registered trade name.

superior in corrosion resistance. Tests should be used to determine which alloy to select for a specific environment.

Hastelloy B-2

Characteristics of Hastelloy B-2 are that it is:

1. An improved wrought version of Hastelloy B and has the same corrosion resistance but has improved resistance to knife-line attack.
2. Resistant to intergranular attack, thus making it suitable for most applications in the as-welded condition. It has excellent resistance to pitting and stress corrosion cracking.
3. Resistant to HCl gas, acetic, and phosphoric acids.
4. One of the few metallic materials that can be used for hydrochloric acid service. Alloy B-2 shows good corrosion resistance even to strong hot HCl solutions, as long as there are no strong oxidizing agents present, such as ferric or cupric chloride. Oxidizing ions as low as 50 parts per million can degrade the corrosion resistance of this alloy. (Ferric or cupric salts may develop when HCl comes in contact with iron or copper.)
5. Resistant to boiling sulfuric acid solutions up to about 60% concentration. (In most cases, however, lower cost materials are available for this service.)

Hastelloy C-276

Characteristics of Hastelloy C-276 are that it is:

1. An improved wrought version of Hastelloy C. It has the same corrosion resistance as Hastelloy C with improved fabricability.
2. One of the few materials that will resist the corrosive effects of hypochlorite and chlorine dioxide solutions.
3. Extremely corrosion-resistant to strong oxidizers such as ferric and cupric chlorides or hypochlorites and to wet chlorine gas at room temperature, but attack becomes appreciable in hot solutions, especially in higher concentrations.
4. Resistant to hot contaminated acids, solvents, formic acid, acetic acid, and acetic anhydride.

Hastelloy G-3

Characteristics of Hastelloy G-3 are that it is:

1. An improved wrought version of Hastelloy G. It has the same general corrosion resistance as Hastelloy G with improved resistance to intergranular attack and improved weldability.
2. Very resistant to mixed acids.
3. Very resistant to hot sulfuric and phosphoric acids in the as-welded condition.
4. Tolerant of the corrosive effects of both oxidizing and reducing agents and can handle both acid and alkaline solutions.
5. Applied principally in hot concentrated sulfuric, hydrofluoric, and fluorosilic acids involved in the manufacture of hydrofluoric acid.

Hastelloy G-30

Characteristics of Hastelloy G-30 are that it:

1. Has superior corrosion resistance to phosphoric acid compared to most other nickel and iron-base alloys.
2. Resists highly oxidizing acids such as nitric/hydrochloric, nitric/hydrofluoric, and sulfuric acids.
3. Can be used in the as-welded condition for most process applications.
4. Has been found satisfactory for many applications, including pickling acid operations, petrochemicals, fertilizer manufacturing, pesticide manufacturing, and gold ore extraction.

Hastelloy C-22

Characteristics of Hastelloy C-22 are that it:

1. Is a versatile nickel-chromium-molybdenum alloy claimed to have better overall corrosion resistance than other NiCrMo alloys available today.
2. Has outstanding resistance to pitting, crevice corrosion, and stress corrosion cracking.
3. Has excellent resistance to oxidizing aqueous media, including wet chlorine and mixtures containing nitric acid or oxidizing acids with chloride ions. It is also resistant to reducing aqueous media.
4. Is resistant to strong oxidizers such as ferric and cupric chlorides.
5. Is resistant to chlorine, formic, and acetic acids, and acetic anhydride.
6. Can be used in the as-welded condition without heat treatment.

□ 5.4 Aluminum and Aluminum Alloys

There are many wrought and cast aluminum alloys. The ones most commonly used in corrosive service are shown in Table 5.9. For chemical service, they have the advantage over some other metals by having colorless and nontoxic corrosion products; therefore, fluids and solids handled in aluminum equipment are not discolored by the aluminum. Aluminum alloys do not become brittle at subzero temperatures like various other metals. Aluminum alloys rapidly lose strength at temperatures of 350 F (177 C) or higher.

TABLE 5.9 — Nominal Compositions and Various Uses of Aluminum Alloys

Alloy No.	Mn	Mg	Cr	Cu	Si	Al	Various Uses
1060						99.6%	Railroad tank cars and chemical equipment; premium grade used in special applications, including shipping drums and handling equipment for hydrogen peroxide where impurities catalyze peroxide decomposition.
1100						99.0	Commercial "pure" grade; used for acetic acid service tanks, pumps, and handling equipment.
3003	1.2					bal	About the same corrosion resistance as AA 1100 but with higher mechanical properties; used in tanks, shipping containers, and heat exchangers.
5052 5083	 0.7	2.5 4.5	0.25			bal bal	Used for equipment requiring higher strengths (28-46 ksi, depending on temper), such as hopper cars, chemical truck bodies, where minimum wall thicknesses are required, and corrosive conditions are mild. Unfired pressure vessels; Alloy 5083 cannot be used in the strain-hardened condition over 150 F, because it will fail by SCC in mildly corrosive environments.
6061		1.0	0.25		0.5	bal	Bars and shapes for structural purposes where corrosion is mild.
355 356		0.5 0.3		1.3	5.0 7.0	bal bal	Cast alloys; these alloys can be heat treated to good mechanical properties (25-39 ksi, depending on temper and alloy); corrosion resistance generally the same as the higher strength wrought aluminum materials; marine hardware, wheels, axle housings.

Corrosion Resistance of Aluminum and Aluminum Alloys

Pure aluminum has the greater corrosion resistance when compared to aluminum alloys. As alloying elements are added to pure aluminum, the corrosion resistance is somewhat reduced while the mechanical properties are improved. For this reason, when strength and high corrosion resistance both are required, the designer can specify

Alclad† aluminum alloys, which are aluminum structural alloys covered with a thin skin of pure aluminum.

Aluminum and aluminum alloys are corrosion resistant because of their protective oxide film. They are satisfactory materials of construction only in those services where this film can be maintained. *Anodizing* is a treatment which enhances the corrosion resistance and wear resistance of aluminum alloys by greatly increasing the aluminum oxide (Al_2O_3) thickness on the surface. Aluminum is amphoteric and susceptible to attack in strongly alkaline, as well as strongly acidic, solutions. The protective film is only stable in a pH range between 4.5 and 8.5. Corrosion resistance in neutral and many acid conditions is good.

Water. Aluminum is used for handling distilled water and for the distillation of high purity water. Since most plant cooling waters contain impurities, including heavy metal salts that can cause excessive pitting, aluminum and aluminum alloys cannot be used unless the water is inhibited. Copper and mercury salts are particularly dangerous to aluminum, as are other heavy metals. Aluminum and aluminum alloys have excellent resistance to steam and steam condensate, provided there are no entrained alkaline boiler compounds present.

Nitric acid. Aluminum is not sufficiently resistant to mineral acids to allow use except for strong nitric acid, which, at low temperatures, will not break down the protective oxide film and hence good corrosion resistance is permitted. US Interstate Commerce Commission (ICC) regulations allow the shipment of nitric acid in aluminum containers at concentrations above 82%. Nevertheless, care must be exercised by the designer in using aluminum for this service because if concentrations of the nitric acid go below 82%, aluminum will show an increased corrosion rate. Aluminum alloys are very resistant to ammonium nitrate solutions, as long as there are no traces of copper, tin, or lead present. Above 92% nitric acid, the aluminum has greater corrosion resistance than AISI 304L. However, the *welding deposit* must be AA 1100 alloy to resist the corrosion of the acid.

Organic acids and compounds. Aluminum alloys are used throughout industry to handle organic acids including propionic and acetic acids. These acids are regularly shipped in aluminum tank cars. The anhydrides of organic acids are also handled well in aluminum. When using aluminum for these applications, care must be exercised because only small changes in conditions can greatly change performance. For instance, aluminum shows excellent resistance to acetic acid at room temperature at any concentration. Rates are still relatively low up to 120 F (49 C). At the boiling point, it is badly attacked at low

---

†Registered trade name.

and intermediate concentrations, but corrosion rates decrease rapidly about 90% and resistance to glacial acid is good.

☐ 5.5 Carbon and Low Alloy Steels

Many thousands of tons of steel are used every year by the chemical industry and other industries. This may seem paradoxical because of the poor reputation of steel for corrosion resistance since it rusts in water and in the atmosphere. However, because it is relatively inexpensive and easy to weld, carbon and low alloy steels are used widely and successfully for numerous process conditions, alkaline solutions, and for certain acids. There are a few exceptions, but in general, the corrosion resistance of annealed carbon steels are all about the same; however, welding and heat treatment can produce phase changes that can cause changes in the corrosion resistance.

The corrosion resistance of the various alloys listed in Tables 5.10 and 5.11 are predicated more on the condition of the steel as opposed to actual chemical compositions. The quality of the surface, such as

TABLE 5.10 — Nominal Compositions and Various Uses of Low Carbon Steel

ANSI/ASTM Specification	Nominal Composition (bal Fe)	Various Uses
A-36	C	Covers structural steel shapes and lists other specifications covering specific mill shapes.
A-283	C	Lowest cost plate material made to ASTM specifications. It is a general purpose steel and used where a minimum amount of quality control testing is acceptable. Its use is limited to a maximum temperature of 650 F for ASME Code vessel use.
A-285	C	Differs from A-283 primarily in the amount of quality control testing that is done. The ASME code permits its use to a maximum temperature of 900 F. Perhaps the most widely used carbon steel for moderate temperature and pressure service.
A-299	C-Mn-Si	Contains 1.25% manganese, which results in the highest strength range of any of the carbon steel plate materials. The addition of manganese does not materially increase corrosion resistance.
A-515	C-Si	This is a lower grade steel than A-516 with an upper temperature limit of 850 F; not used for low-temperature service.
A-516	C-Si	This silicon deoxidized steel plate is designated by ASME to be used at temperatures up to 1000 F; used for low-temperature service down to −20 F because of fine grain size, which imparts excellent low-temperature−impact properties.

TABLE 5.11 — Nominal Composition and Uses of Low-Alloy Steels

ANSI/ASTM Specification	Nominal Composition	Various
A-202B	1/2Cr-1Mn	All the low-alloy steels listed have greater strength than is available in plain low carbon steels. The grades for the various specification numbers differ primarily in the type and amount of alloying elements that are present, which, in turn, provides differences in mechanical properties. These steels are more expensive than carbon steels both in price per pound and cost of fabrication. The cost is 50 to 100% greater than carbon steels. Additionally, if equipment must be fabricated to ASME Code rules, the requirements for stress relief and radiographic examination are more restricted. Used for a myriad of structural applications.
A-204C	C-1/2Mn	
A-302B	Mn-1/2Mo	
A-387B	1Cr-1/2Mo[1]	
A-387D	2 1/4Cr-1Mo[1]	
A-533	1 1/4Cr-1/2Mo[1]	

[1] In addition, CrMo steels are used in applications where more resistance to oxidation from air, oxidation from steam, high temperature hydrogen attack, and graphitization is required.

rough or smooth, many or few inclusions, the adherence of mill scale, the presence of locked-up stresses, etc., directly affect the steel's resistance to corrosion. Therefore, the specification of a particular carbon steel by the designer will depend more on inherent mechanical properties (either alone or by heat treatment) that the steel possesses rather than for its corrosion resistance. Of course, the quality of the steel must be carefully specified.

Alkaline Solutions

Steel is generally resistant to a wide range of alkaline solutions. It is used for the storage and handling of such solutions as anhydrous and aqueous ammonia and sodium hydroxide. Sensitivity to oxygen is practically nonexistent at pH values above 10. Stress corrosion cracking may occur in and adjacent to unstressed relieved welds in carbon steel equipment containing sodium hydroxide solutions if the temperature is well above ambient. See Chapter 3.8, Table 3.4 and Figure 3.55 for detailed information. Inhibitors are frequently used to allow carbon steels to be used in corrosive waters.

Sulfuric Acid

Carbon steel use. Throughout the world, more sulfuric acid (H_2SO_4) is produced than any other inorganic acid. It is generally considered the most important industrial chemical.[42] Carbon steel has long been used as the material of construction for handling and storage of this acid at low temperatures.

Because of the wide variation in the corrosion resistance of carbon steel in sulfuric acid, this section has been purposely expanded so that the designer will understand the factors that control this variation and will know what to do to avoid excessive corrosion.

The wide use of carbon steel in sulfuric acid service is based on its low cost and ease of fabrication coupled with its normally satisfactory corrosion resistance. However, the designer should recognize that acid concentration, temperature, and velocity can have a significant effect on corrosion rates. Before carbon steel is selected as a material of construction, the designer should conduct a thorough study of the actual exposure conditions. In some designs, stainless steels, high-nickel alloys, protective coatings, or anodic protection of carbon steel may be required.

Variables. Carbon steel owes its corrosion resistance to the formation of a film of ferrous sulfate, which is a product of corrosion that forms on the steel. If this protective layer is *not disturbed*, the acid velocity remains low, the acid is *not diluted*, and the *temperature is not raised*, the corrosion rates will be consistently low. The corrosion rates at ambient temperatures from 65 to 96% acid strength are 5 to 20 mils per year and 0 to 5 mils per year from 96 to 110%. (See Figure 3.81.) Reliance on such corrosion rates may prove hazardous for the designer if he does not keep these significant variables in mind. Corrosion penetration rates can be higher than 200 mils per year, depending on these variables.

Acid concentration. The concentration of sulfuric acid is also a dominant factor in the corrosion resistance of carbon steel. Below 65% acid concentration, even at ambient temperatures, carbon steel corrodes. As the concentration decreases to below 50%, the ferrous sulfate corrosion product rapidly goes into solution, thus fostering very high corrosion rates.

Sulfuric acid is hygroscopic and can readily pick up moisture from the air if precautions are not taken. Such dilution of the acid can lower its concentration and lead to aggressive corrosion as mentioned above. An example of this was shown in Chapter 4.4, where a tank bottom corroded through because of acid dilution. Provision must be made by the designer not to allow moisture filled air to be drawn into the sulfuric acid tank during fluctuations in temperature.

Carbon steel should never be used for service in sulfuric acid much under 65% concentration because corrosion rates can be up to 50 mils per year at ambient temperatures and over 200 mils per year at temperatures greater than 175 F (79 C).

Temperature. The effect of temperatures is clear. Carbon steel is not suitable for use at temperatures much over ambient, according to Figure 3.81. Actually, few materials can withstand the 60 to 98% sulfuric acid at temperatures over 250 F (121 C), with the exceptions of tantalum, acid brick-lined steel, platinum, and fluorocarbon plastic.[44]

Ferrous sulfate protective film. When the ferrous sulfate protective film is disturbed by hydrogen bubbles (generated from the corrosive reaction of carbon steel and sulfuric acid), flowing in a steady stream over specific paths, hydrogen grooving can result. (See Chapter 3.13 for further information on hydrogen grooving.)

Velocity. The effect of velocity on the corrosion rates of sulfuric acid on carbon steel are discussed in Chapter 3.4. Several examples are shown where carbon steel tolerated low sulfuric acid velocities, but not higher velocities.

What the Designer Can Do to Reduce Excessive Corrosion of Carbon Steel in Strong Sulfuric Acid

There are several steps a designer can take to reduce excessive corrosion of carbon steel in strong sulfuric acid, including:

Select pipe diameters that will keep the velocity low. When designing pipelines, select pipe diameters that will keep the velocity as low as is economically possible. Try to keep velocities at 8 feet per second or below.

Avoid turbulence whenever possible. Turbulence should be avoided whenever possible because it constitutes local high-velocity acid. Use wide sweep elbows. Where sharp bends must be made, use heavier fittings to afford a greater corrosion allowance.

Specify that outside surfaces of sulfuric acid tanks be painted white. Because acid temperature is critical, the designer should specify that sulfuric acid storage tanks be painted white to reflect the solar rays. Tanks may also be thermally insulated for this purpose.

Obtain an analysis of the sulfuric acid that is to be stored or handled. The designer should obtain an analysis of the sulfuric acid to be stored or handled. If the iron content is below 30 ppm, the acid will tend to pick up more iron. This will increase the corrosion rate. When this is the case, the designer should increase the corrosion allowance.

Require a dryer in conjunction with the vent tube to prevent acid dilution. To keep the acid from being diluted as air in drawn into a tank as the acid is being emptied, a dryer may be required in conjunction with the vent tube. All manway doors and other openings should be closed tightly to exclude air. This should be included in the designer's overall materials guide.

Specify that all welds be of high quality. All welds in sulfuric service should be of high quality, showing full penetration. All butt welds, when size permits, should be double, i.e., welded from both inside and out. Piping with diameters too small to permit double butt welds should be welded using consumable insert rings. (See Chapter 3.1 for further information on rings.) All fillet welds inside tanks must be full penetration, as noted in Chapter 4.1 with the example of barge welds.

Specify multipass welding when using manual arc welding. When manual metal arc welding is to be used, all welding should be specified to be multipass. The reason for this is that in a weld made with a single pass, there is more of a chance that slag will be distributed in a path through the depth of the weld than there would be if more than one pass is made. Sulfuric acid readily attacks such slag leaving voids in the weld deposit causing it to be unsound.

> *Example* — The inside of a carbon steel horizontal tank was inspected that had held 66 degree Baume (93%) sulfuric acid for many years. The shell weld seams had been properly designed using double butt welds. However, it was reported that the welding had been accomplished by using 1/4-in. (7-mm) electrodes (AWS E-7010). The butt weld had filled the welding groove in one pass. The inside weld was found to be riddled with holes, some to the bottom of the weld pass. Pieces of the weld could be literally lifted out of the weld groove. To make repairs, the inside weld was completely ground out and welded in three passes with a 5/32-in. (4-mm)-diameter electrode (AWS E-6010). No further weld corrosion was reported.

Specify the removal of the mill scale from the sheet or plate ordered. In the purchase order for steel, the designer should specify that the mill scale must be removed from the sheet or plate ordered. Mill scale is readily consumed by sulfuric acid, leaving pock marks that can become sites for localized corrosion.

Make sure weld surfaces are smooth. Welds with rough surfaces should be lightly ground or sandblasted to obtain a smooth surface. It is not necessary to grind flush. The welds should be smooth enough so that acid cannot be retained on the surface after the acid is emptied.

Use standards for guidelines. As a guideline, the designer can use a corrosion allowance of 1/4 in. (7 mm) for storing 70 to 100.5% sulfuric acid and 1/8 in. (3 mm) for 100.5 to 110% acid, providing the temperature is ambient. Use API Standard 650 (1980) as a guide when designing sulfuric acid tanks. Generally, ASTM Standard A-516 Grade 70 or A-662 Grade B steel will satisfy the requirements as long as the notch toughness properties are met.

Consider anodic protection for critical applications. For critical applications of carbon steel tanks in sulfuric acid service, the designer has the option of specifying anodic protection. (Refer to Chapter 6.4.) Anodic protection[42] can lower mild steel corrosion rates to 1 to 2 mils per year, which is considerably lower than corrosion rates would be without such protection. Anodic systems have been satisfactory for controlling corrosion in many sulfuric acid tanks and other process equipment.

Other Acids

Carbon steel has very poor resistance to hydrochloric, nitric, phosphoric, and acetic acids and should not be used. However, there are mixed acids for which carbon steel is a permissible material of construction. For example, certain mixtures of sulfuric acid, nitric acid, and water can be contained in steel.

Atmosphere

In all atmospheres (except perhaps under very dry conditions), steel surfaces must be painted for protection. (See Chapter 6.6.) Industrial and marine atmospheres are the most aggressive to steel. Low alloy steels containing small amounts of copper (1/4%), such as Corten[†] steel, may show better resistance than plain carbon steel. This is attributed to the formation of a more adherent rust or oxide film that minimizes further rusting. It is difficult to predict the corrosion resistance of carbon and low allow steels because the atmospheric conditions in various locations vary greatly.

Water

The corrosion resistance of steel in various waters depends on many factors, such as pH, amount of dissolved solids, amount of dissolved oxygen, temperature, and velocity. Hydrogen evolution from carbon steel in water can occur at pH values as high as 4.5 and tends to decrease as the pH value is increased. However, even at pH levels of 7 to 10, care must be exercised to obviate hydrogen embrittlement.

[†]Registered trade name.

The protective films formed at lower pH values are often discontinuous and hence not very protective. At pH values of 10 or above, a tightly adherent film that is very resistant to corrosion forms. Inhibitors are frequently used to allow carbon steels to be used in corrosive waters.

Steam Generation

Carbon steel is used universally for steam generating equipment, pipelines, and other handling equipment. Excellent service life is obtained if the boiler water is properly deaerated and/or chemically treated. Sulfites, hydrazine, and other chemicals are used to scavenge the oxygen in the water; otherwise, the dissolved oxygen would cause aggressive attack on the steel.

Inorganic Salt Solutions

Solutions of acidic salts, such as NH_4Cl and $AlCl_3$, are very corrosive to steel. Alkaline salts, such as Na_2CO_3, however, are generally noncorrosive except for strongly oxidizing salts like sodium hypochlorite. When brines are handled in steel, they should be kept at pH levels above 7 and should be inhibited or deaerated.

High-Temperature Corrosion

The oxide coating that forms on steel is protective up to about 1,000 F (593 C). High-temperature hydrogen attack increases with temperature and the hydrogen partial pressure. Alloying ingredients in steel significantly increase resistance to high-temperature corrosion. Alloying elements, such as chromium and molybdenum, are used.

Gases

Alloy steels are used to handle gases containing carbon monoxide and in ammonia mixtures, they are used for nitriding steel.

☐ 5.6 Copper and Copper Alloys

Copper and copper-base alloys have been extensively used throughout industry for corrosion resistance. Table 5.12 shows 15 copper alloys that are generally most used for this purpose. The wrought alloys are very easily formed because of their excellent ductility. The cast alloys have excellent castability. The welding of some alloys is complicated by high thermal conductivity; however, the exception is high silicon Bronze A, which has a much lower thermal conductivity and can consequently be welded with the insert gas process, almost as easily as stainless steel.

TABLE 5.12 — Nominal Compositions and Various Uses of Copper-Base Alloys

Copper Development Association No.	Name	Cu (%)	Zn (%)	Sn (%)	Ni (%)	Others (%)	Various Uses
122	Phosphorous Deoxidized Copper	99.9 minimum				P-0.015 to 0.040	Plumbing and gas lines; gasoline and water lines; heat exchanger tubes.
230	Red Brass	85	15				Heat exchanger tubes, flexible hose.
270	Yellow Brass	65	35				Springs, plumbing accessories.
443	Admiralty Brass	71	28	1			Condenser, heat exchangers, evaporator tubes, condenser tube sheets, distillation tubes, ferrules.
465	Naval Brass	60	39	1			Marine hardware, nuts, valve stems, condenser plates, bolts.
687	Aluminum Brass	77	21			Al-2	Heat exchanger, evaporator and condenser tubing, distillation tubes.
521	Phosphor Bronze B	92		8		P-0.25	Textile machinery, bridge bearing plates, chemical hardware, bellows, springs.
614	Aluminum Bronze D	91				Al-7 Fe-2	Marine sheeting, tanks, nuts, bolts, condenser tubes.
655	High Silicon Bronze A	94	1		0.5	Si-3 Mn-1 Fe-0.5	Seamless tubes, heat exchangers, welded hot water storage heaters, fasteners.
706	90-10 Cupro-Nickel	88			10		Salt water piping, evaporator tubes, chemical process equipment, distillation tubes, condenser tubes and tube sheets, heat exchanger tubes.
710	80-20 Cupro-Nickel	78			20		
715	70-30 Cupro-Nickel	69			30		
905	Tin Bronze Casting	88	2	10			Steam fittings, pump impellers, pump bodies, valves, bushings, bearings, piston rings.
836	Leaded Red Brass Casting	85	5	5		Pb-5	Pipe fittings, plumbing goods, low-pressure valve bodies.

Corrosion Resistance of Copper and Copper Alloys

Referring to Table 5.12, phosphorous deoxidized copper is the first material listed and is universally used as tubing. The next five alloys, mainly comprising copper and zinc, are members of the *brass* family. These alloys are used widely in industry for tubing and process equipment. As described in Chapter 3.12, these alloys are susceptible to selective leaching or dezincification called *dealloying*, especially in hot, polluted, or mildly acid solutions. The corrosion resistance can be increased by adding inhibiting elements, such as phosphorus, antimony, or arsenic to the copper melt. These alloys are also susceptible to stress corrosion cracking, as noted in Chapter 3.8. The addition of inhibiting elements does not reduce the stress corrosion cracking tendency. The alloys listed as *bronzes* in Table 5.12 are essentially composed of copper and tin and have a corrosion resistance much greater than the brasses. They are not susceptible to dealloying and have increased corrosion resistance, but they do not have immunity to stress corrosion cracking. They are also more expensive than the brasses. The three *cupro-nickels* listed have outstanding corrosion resistance and are widely used for salt water service. The last two alloys listed are *copper alloy castings*. They are very useful in industry for intricate parts such as valve bodies and pipe fittings.

Basic solutions. Solutions of ammonium hydroxide are very corrosive to copper and its alloys because of the formation of a soluble complex copper-ammonia compound. Hot dilute caustic solutions are only mildly corrosive to the bronzes.

Mercury. Mercury and mercury compounds will rapidly crack stressed copper alloy parts; in fact, the mercurous nitrate test is a regularly used laboratory test to detect residual stresses in copper alloys. Mercury will even crack elemental copper.

Neutral and alkaline salts. Copper and its alloys are generally resistant to neutral or alkaline salts solutions; however, there should be definite assurance that there is no acid involved.

Oxidizing salts. Solutions of oxidizing salts, such as hypochlorite, are damaging to copper and copper alloys. Acid salts and oxidizing acid salts, such as ferric chlorides, are also very corrosive as are all oxidizing acids.

Organic compounds. Many organic compounds, such as esters, alcohols, ketones, etc., are handled in copper equipment. Deaeration is not required unless appreciable acidity is developed.

Non-oxidizing acids. Copper and copper-base alloys are quite resistant to dilute solutions of most non-oxidizing acids, either hot or cold, and also cold concentrated solutions, provided that the oxygen is completely excluded by nitrogen blanketing or assurance of reducing conditions. Copper and copper-base alloys, in direct contrast to austenitic stainless steels, thrive on reducing conditions but fail in oxidizing conditions. Copper is actually a noble material and consequently, hydrogen does not generally evolve during corrosion. For this reason, it is not attacked by acids unless oxygen or oxidizing agents are present. In such cases, the corrosion mechanism involves the reduction of oxygen at the cathode. In some applications in industry where stainless steels would not be satisfactory and when only very expensive high alloys would be satisfactory, the cupro-nickels and aluminum bronzes protected from oxygen by nitrogen blankets have given satisfactory service with great monetary savings.

Nitric acid. Nitric acid, which is an oxidizing acid, rapidly attacks copper and copper-base alloys.

Acetylene. In some cases, the mild corrosion of copper and copper alloys releases undesirable corrosion products. The copper salts formed may discolor the finished product or act as a catalyst for undesirable side reactions. For example, acetylene forms a very explosive compound with silver and copper alloys when moisture and alloys containing more than 65% copper interact. This interaction may cause an explosion. Therefore, wet acetylene under pressure should never contact even one copper-base fitting or part (or silver-soldered part).

Marine use. Copper compounds are highly toxic to most organisms. For instance, copper was once used extensively for sheathing the bottom of wooden ships to prevent fouling caused by the buildup of marine organisms. Copper in paints can prevent or slow down the buildup of marine organisms that can cause localized corrosion on steel ship bottoms. Brass, aluminum bronze, cupro-nickels, tin-bronze castings, and other copper-base alloys are used extensively for hardware and sheathing for use in ships and other marine environments.

☐ 5.7 Titanium and Titanium Alloys

There are four grades of titanium and titanium alloys that are used extensively in industry. (See Table 5.13.) These alloys are corrosion resistant because of their protective oxide films and are best under oxidizing conditions similar to austenitic stainless steels. The strength to weight ratio of titanium is very high. Titanium is easily welded, but the

TABLE 5.13 — Nominal Compositions and Various Uses of Titanium and Titanium Alloys

ASTM B-265 (Grade)	C (maximum)	Fe (maximum)	Al	O	V	N	Pd	Minimum Tensile (Strength/ksi)	Various Uses
2	0.10	0.30				0.03		50	Uses include parts requiring excellent corrosion resistance, high formability, good strength to weight ratio, and good weldability; plate, sheet, strip, and tubing available; marine hardware aircraft parts, ordnance parts, parts and hardware for the space industry and tubing parts, vessels, heat exchangers, etc. for the chemical industry. The latter uses are based primarily on resistance to chloride stress corrosion cracking, for which the additional cost can be justified.
3	0.10	0.30				0.05		65	
5	0.20	0.40	6	0.25	4			130	
7	0.10	0.30				0.03	0.12 to 0.15	50	

molten metal must be completely shielded from the oxygen and nitrogen of the air. (See AWS specification D10k, "Practices and Procedures for Welding Titanium Pipe and Tubing" for further details on welding.) The use of titanium in industry is limited to a maximum temperature of 600 F (315 C). Commercially pure titanium is available in several grades. Grade 2 is most commonly used for industrial equipment, since it is approved by the ASME Code for Unfired Pressure Vessels. Grade 7, containing small amounts of palladium, offers enhanced corrosion resistance in certain environments. Grades 3 and 5 are used where increased tensile strength is required (65,000 and 130,000 psi/min, respectively).

Corrosion Resistance of Titanium and Titanium Alloys

Depending on the environment, titanium and its alloys have varied degrees of corrosion resistance.

Brackish water and seawater. Titanium has excellent resistance to corrosive attack by seawater and most chloride salt solutions. It is not susceptible to stress corrosion cracking in plant cooling waters or chloride solutions like austenitic stainless steel.

Titanium is susceptible to crevice corrosion in solutions that the metal would ordinarily be resistant to. Of the four titanium grades listed in Table 5.13, Grade 7 is much more resistant to this form of attack and yet is not entirely immune.

Hydrochloric and sulfuric acids. Titanium has poor resistance to reducing mineral acids, such as sulfuric and hydrochloric. However, if strong oxidizing agents are present, such as ferric or cupric salts, titanium can be used with fairly low corrosion rates. Grade 7 has substantially better resistance under non-oxidizing or slight reducing conditions than Grades 2 or 3.

Hydrofluoric acid. Titanium is vigorously attacked by hydrofluoric acid and by any solution that contains more than 5 to 50 ppm of fluoride ions.

Organic acids. This material is very resistant to some organic acids. It is resistant to acetic acid as long as there is around 0.2% water present. Oxalic, trichloroacetic, and formic acids corrode titanium.

Nitric acid. Titanium has good resistance to nitric acid under certain conditions. It is very resistant to dilute and concentrated nitric acid. At temperatures above atmospheric boiling, it has much greater corrosion resistance than austenitic stainless steels.

Oxidizing salts. Titanium and titanium alloys have excellent resistance to oxidizing salts such as hypochlorites, cupric, and ferric chlorides. These salts tend to pit most other metal but actually inhibit corrosion in titanium.

Wet chlorine. Titanium has excellent resistance to wet chlorine, except where there are crevices which will cause rapid failure. Dry chlorine can cause a violent exothermic reaction. At least 1% water must be present to prevent this reaction.

Red fuming nitric acid. Titanium is catastrophically attacked in red fuming nitric acid with high NO_2 content and low water content.

Galvanic corrosion effects. Because titanium has the ability to easily and immediately passivate, it is seldom affected by galvanic corrosion.

☐ 5.8 Unified Numbering System (UNS)

During this century, there has developed a bewildering array of metals and alloys, each with its own designation or number. Because there was not a central organization established at that time, these designations have come from trade names, trade associations, and technical organizations; consequently, there was confusion regarding metal and alloy identities. The composition of type 304 stainless steel, for instance, was different in ASTM, SAE, and AISI standards. The differences were small, but there were discrepancies. There were also instances when the same number was assigned to two entirely different alloys and other cases when there was more than one identification number for the same material. Starting in 1969, SAE and ASTM conducted a study for the Army Materials and Mechanics Research Center, Watertown, Massachusetts, to determine the feasibility of developing a new consistent numbering system for metals and alloys.

Out of the SAE/ASTM study evolved the Unified Numbering System (UNS), which has eliminated the confusion in numbering metals and alloys. The designer is cautioned that a UNS number by itself is not a specification. It is a number carefully controlled by SAE and ASTM in conjunction with various technical societies to indicate established specifications published elsewhere, which limits the form, condition, quality, chemical analysis, etc., of various metals and alloys.

Whenever practical, the original number designation of a specific metal or alloy is incorporated in the UNS number. For instance, AISI 304 stainless steel is UNS S 30400; Nickel Alloy 600 is UNS N 06600; and Aluminum Alloy 6061 is UNS A 96061.

The primary series of UNS numbers are shown in Table 5.14. The UNS numbers corresponding to all the metals and alloys shown in

TABLE 5.14 — Primary Series of UNS Numbers[1]

UNS Series	Metal
Nonferrous metals and alloys	
A00001-A99999	Aluminum and aluminum alloys
C00001-C99999	Copper and copper alloys
N00001-N99999	Nickel and nickel alloys
R00001-R99999	Reactive and refractory metals and alloys
Ferrous metals and alloys	
J00001-J99999	Cast steels
K00001-K99999	Miscellaneous steels and ferrous alloys
S00001-S99999	Heat and corrosion resistant (stainless) steels

[1] Abbreviated table from SAE HS1086a and ASTM DS-56A.

Tables 5.2 through 5.13 are shown in Table 5.15. As mentioned before, the designer should become familiar with this system because it will be used more and more and should help the designer in assuring that he actually receives the corrosion-resistant materials that he requires. The designer should refer to "Unified System for Metals and Alloys," SAE Standard HS1086a and ASTM Standard DS-56A for a compilation of all UNS numbers assigned so far.

☐ 5.9 Overall Materials Guide (OMG)

The proper selection of materials of construction is very important, as emphasized before. If the optimum material is not determined, either the equipment may fail or the cost of the equipment will be higher than necessary. Past experience shows that after a plant has been operating, it is sometimes difficult to determine what materials of construction were used for a piece of equipment without laboriously studying blue print files, specifications, and purchase orders. In addition, the reasons why certain materials were selected are often obscure or unknown. The purpose of an Overall Materials Guide (OMG) is to list in one document the materials of construction for each piece of equipment in a specific plant, and the reasons why these specific materials of construction were chosen.

TABLE 5.15 — Unified Numbering System

Refer to Table No.	Type of Metal or Alloy	Designation No.	UNS No.
5.2	Austenitic Stainless Steel	AISI 301	S30100
		AISI 302	S30200
		AISI 304	S30400
		AISI 304L	S30403
		AISI 321	S32100
		AISI 347	S34700
		AISI 303	S30300
		AISI 305	S30500
		AISI 308	S30800
		AISI 308L	S30803
		AISI 309	S30900
		AISI 310	S31000
		AISI 309Cb	S30904
		AISI 316	S31600
		AISI 317	S31700
		AISI 316L	S31603
		AISI 201	S20100
		AISI 202	S20200
		AISI 204	S20400
		AISI 204L	S20403
5.3	Martensitic Stainless Steel	AISI 403	S40300
		AISI 410	S41000
		AISI 414	S41400
		AISI 416	S41600
		AISI 420	S42000
		AISI 431	S43100
		AISI 440A	S44002
		AISI 440B	S44003
		AISI 440C	S44004
5.4	Ferritic Stainless Steel	AISI 405	S40500
		AISI 431	S43100
		AISI 442	S44200
		AISI 446	S44600
5.5	Precipitation-Hardening Stainless Steel	17 to 4 PH	S17400
		17 to 7 PH	S17700
		PH15-17Mo	S15700
		Stainless W	S17600
5.7	Cast Stainless Steel	CA-15	J91150
		CA-40	J91153
		CA-50	J92615
		CF-8	J92600
		CF-8M	J92900
		CF-20	J92602
		CF-8c	J93400
		CF-16f	J92701
		CD-4MCu	J93402
		CH-20	J93402
		CK-20	J94202
		CN-7M	J95150
		HC	J92605
		HF	J92603
		HH	J93503
		HK	J94224
5.8	Nickel and Nickel Alloys	200	N02200
		201	N02201
		400	N04400
		600	N06600
		625	N06625
		800	N8800
		20 Cb-3	N8020
		825	N8825

TABLE 5.15 — Unified Numbering System (continued)

Refer to Table No.	Type of Metal or Alloy	Designation No.[1]	UNS No.
		Hastelloy B-2	N10665
		Hastelloy C-276	N10276
		Hastelloy G-3	N06007
5.9	Aluminum and Aluminum Alloys	AA 1060	A91060
		AA 1100	A91100
		AA 3003	A93003
		AA 5052	A95052
		AA 5083	A95083
		AA 6061	A96061
		AA 355	A33550
		AA 356	A13560
5.10	Low Carbon Steel	ASTM A-36 (Shapes)	K02600
		ASTM A-283	K02800
		ASTM A-285 (c)	K02801
		ASTM A-299	K02803
		ASTM A-515(65)	K02800
		ASTM A-516(70)	K02700
5.11	Low-Alloy Steels	ASTM A-202B	K12542
		ASTM A-204C	K12320
		ASTM A-302B	K12022
		ASTM A-387B	K12143
		ASTM A-398D	K41545
		ASTM A-533(B)	K12539
5.12	Copper and Copper-Base Alloys	CDA 122	C12200
		CDA 230	C23000
		CDA 270	C27000
		CDA 443	C44300
		CDA 465	C46500
		CDA 687	C68700
		CDA 521	C52100
		CDA 614	C61400
		CDA 655	C65500
		CDA 706	C70600
		CDA 710	C71000
		CDA 715	C71500
		CDA 905	C90500
		CDA 836	C83600
5.13	Titanium and Titanium Alloys	ASTM B-265 Grade 2	R50400
		ASTM B-265 Grade 3	R50550
		ASTM B-265 Grade 5	R56401
		ASTM B-265 Grade 7	R52400

Additional Purposes of the Overall Materials Guide

The OMG serves the following additional purposes:

1. If a materials selection was based on past experience, the OMG will state where that experience was gained (such as at another operating plant.)
2. In a critical piece of equipment, the source and content of the corrosion data backing up the materials choice can be stated.

3. Special precautions can be listed to avoid mistakes during startup. Many corrosion problems can arise because process conditions, such as higher temperatures, can be far different during startup than during subsequent plant operations.
4. The OMG can be turned over to plant operations personnel when the plant goes on stream. Any knowledge of process limitations or precautions regarding corrosion resistance can be noted.
5. The OMG can be invaluable in later years when the plant is contemplating for one reason or another, the substitution of another material of construction for a specific piece of equipment or process.

Overall Materials Guide Example

Table 5.16 shows a complete OMG compiled for a simulated plant. It should be emphasized that an OMG is not intended to supplant the normal materials specifications, purchase orders, equipment lists, or operating procedures.

TABLE 5.16 — Overall Materials Specification (OMG) of a Typical Plant Installation

Equipment Name	Materials of Construction	Reasons for Selection
Special Gas Heater	Tubes: SA213 AISI 304L[1] Shell and Tube Sheets: SA240 AISI 304L[1]	Gases on both sides. Process gas, tube side in at 544 F (284 C) and exits at 346 F (174 C). Other gas, shell side entrance is 212 F (100 C) and exit is 468 F (242 C). These temperatures are well within temperature limitation for AISI 304L, and this alloy will withstand the corrosion of condensate formed during shutdown.
Secondary Condenser	Tubes: Unalloyed titanium SB338 welded Tube Sheets: Detaclad SB265 Grade 1 or 2 titanium on SA516 Grade 70 carbon steel Channel: SA240 AISI 304L[1] Shell: SA516 Grade 70 carbon steel	Process gas enters tubes at 362 F (183 C) and exits at 98 F (37 C). Cold water is on the shell side. Water is 102 F (39 C), out at 147 F (64 C). The temperature is too high for AISI 304L in the presence of chloride bearing water; therefore, titanium tubes must be used. The channel can be AISI 304L stainless steel because the water does not contact it. The tubes are very long, therefore, since the expansion of titanium is less than carbon steel; an expansion joint has been included.
Crossover Body Flange	SA516 Grade 70 (per UG-15 ASME Section VIII) (Originally was SA105) with weld overlay of AISI 309L stainless steel deposit	Flange could be solid AISI 304L but would be too expensive. Use standard forged flange made of 70M T/S steel protected with 1/2-in. (12.7-mm) stainless steel overlay. Sample was submitted that showed carbon content to be below 0.03% carbon on surface. Stainless steel required to protect at startup and to assure no iron oxide being formed. Temperature 520 F (271 C) Corrosion resistance of materials of construction based on experience at Plant No. 21.
Acid Storage	AISI 304L stainless steel[1]	Stainless steel required for 57% maximum acid. The extra low carbon grade specified so tanks can be used in the as-welded condition at ambient temperature. Install corrosion coupons or corrosion probe before startup. Observe at 3-month intervals.
Demineralized Water Storage Tank	AA 3003 Aluminum	Aluminum required for water purity. Watch that no heavy elements contact inside or outside.
Ammonia Primary Evaporator	Shell: SA516 Grade 70 carbon steel Tubes: Incoloy 800	Carbon steel would be satisfactory for this application, however, there can be a fluctuating level that could cause accelerated corrosion; therefore, Incoloy 800 is specified.
Special Condenser	Tubes: 90/10 Cu/Ni Tube Sheets: carbon steel (ASTM A-285, Grade C). Water boxes: carbon steel. Shell and tube support plates: carbon steel ASTM A-285, Grade C.	AISI 304L stainless steel tubes could be used; but, because of lower heat transfer coefficient, they would require more area and cost more. Steam in and out at 147 F (64 C). River water in at 93 F (34 C) ant out at 111 F (44 C). Because river water is used at a high temperature, stainless steel tubes might fail by chloride stress corrosion cracking; therefore, cupro-nickel, which is immune to this type of attack, is used. Install corrosion coupons or probe on river water side.
Double Heater	SA213 AISI 304L stainless steel[1] tubes and tube sheet; SA516 Grade 70 carbon steel shell.	Gas enters at 122 F (50 C) and leaves at 212 F (100 C) on the tube side. High pressure steam is on shell side which condenses. Steam in at 370 F (188 C), out at 367 F (186 C). Since H.P. steam will carry no chlorides, AISI 304L tubes may be used. Care must be exercised during startup so that no chlorides are introduced.
Acid Sump Tank	AISI 304L stainless steel[1]	Stainless steel required because of presence of

TABLE 5.16 — Overall Materials Specification (OMG) of a Typical Plant Installation (continued)

Equipment Name	Materials of Construction	Reasons for Selection
		acid. Standard sump tank. Since the tank is underground, 2 shop coats of bitumastic No. 50 on all external surfaces are required.
Reactor Heater	Tubes: AISI 321H[1] Tube Sheets & Shell: AISI 321[1] Ferrules and Heat Shield: Inconel 600 Expansion Joint Bellows: Incoloy 800	Process gas on tube side at 944 F (507 C) entrance and 734 F (390 C) exit. Both conditions demand excellent high temperature and corrosion resistance. Design pressure 160 psi, design temperature 914 F (490 C). AISI 304L stainless steel limited by Code to maximum of 800 F (427 C). AISI 321 is allowed by the Code to be used at 932 F (499 C) temperature. This alloy is stabilized with titanium and, therefore, can be used in the as-welded condition. Gas enters shell at 460 F (238 C) and exits at 745 F (396 C).
Steam Separator	Shell: SA204 Grade A	Steam inlet temperature 766 F (408 C). Corrosion allowance 1/8 in. (3.2 mm). Design temperature was 800 F (427 C); therefore, carbon steel over the long term might not hold up. A carbon-molybdenum steel was therefore used.
Acid Cooler	AISI 304L stainless steel[1]	Cooling acid from 136 F (58 C) to 116 F (47 C). Water 122 F (48 C) from 88 F (31 C). AISI 304L stainless steel required for corrosion resistance. Install corrosion coupons or corrosison probe observed at 3-month intervals.
Heat Exchanger	Tubes: SA213 AISI 304L[1] Tube Sheets: Detaclad SA240 AISI 304L on SA105 carbon steel Channel: SA240 AISI 304L[1] Shell: SA516 Grade 70	Process gas enters at 406 F (208 C) and exits at 276 F (136 C) on the tube side. On the shell side is boiler feed water condensate. Water in at 212 F (100 C) and out at 347 F (175 C). Such water should have practically no chlorides. The required makeup water (probably 10%) is reported to have less than 1 ppm chlorides. Exchanger designed to operate completely full of water. Under these conditions, AISI 304L stainless steel may be used. Carefully check makeup water for chlorides before addition. Do not add if over 2 ppm.
Condensate Drum	Carbon Steel: SA516 Grade 70	Carbon steel satisfactory as long as there is a corrosion allowance of 1/8 in. (3.2 mm). Place corrosion coupons or probe in drum to ascertain if any pitting develops. Install before startup. Observe at 3-month intervals.
Expansion Joint	Incoloy 800	This material has shown good heat, corrosion, and fatigue resistance in plant operations. Incoloy 800 also has good resistance to corrosion by acid. Temperature 1558 F (848 C). Care must be exercised during startup that 1600 F (871 C) temperature limit is not exceeded.

[1]Should be corrosion evaluated per ASTM Standard A-262 C.

■ 6 — CONTROL TECHNIQUES

☐ 6.1 Quality Control

Questions the Designer Should Ask to Control Quality

A designer who is concerned about corrosion resistance must have some way to control quality; otherwise, he may not receive from the manufacturer the corrosion performance he expects from the equipment he has designed. His quality control program should be outlined in the purchase order. Remember that quality can and should be controlled by the designer.

There are many inspection methods available to the designer, but before he specifies one or more of these methods, he should answer the following crucial questions:

1. How corrosive are the process conditions?
2. How toxic are the process conditions?
3. How susceptible is the material of construction to a specific corrosion form, such as crevice corrosion or stress corrosion cracking?
4. How sensitive is the corrosion resistance of the material of construction to shifts in chemical composition?
5. What joining method is to be used? How sensitive is the corrosion resistance of the material of construction to the method of joining, such as welding?
6. How competent is the fabricator? What reputation does he have for self inspection? Does he use Code qualified welders?
7. Is heat treatment required (either for equipment stability or corrosion resistance)?
8. If heat treatment is required, how sensitive are the materials of construction to the heat treatment?
9. How sensitive was the material of construction to mill operations when it was originally produced?
10. If welding is to be the joining method, how important is the filler metal to corrosion performance?

Based on the answers to these questions, a quality assurance program can be formulated.

As broadly defined by the American Society for Quality Control, *quality* is the totality of features and characteristics of a product or a service that depends on its ability to satisfy a given need. This can be briefly stated as *fitness for use*.

Fitness for Use

The designer should decide what inspection or qualification methods are required to assure *fitness for use*. With this in mind, the designer should confer with his inspection department, his technical people, and the potential fabricators to determine which inspection methods should be specified to assure quality. Inspection costs money, so care should be taken not to over-inspect, such as on routine jobs being done by proven competent fabricators. However, inspection under other conditions can be thoroughly justified because of substantial cost savings and elimination of safety hazards.

Inspection Methods

Various inspection methods available to the designer for corrosion control are:

1. Chemical analysis
2. Leak testing
3. Metallographic examination
4. Acceptance testing, such as:

    — Huey Test
    — Strauss Test
    — Test for Heat-Treatable Aluminum Alloys

5. Crack, crevice, and flaw detection, such as:

    — Liquid Penetrant Inspection
    — Magnetic Particle Inspection
    — Radiographic Inspection
    — Ultrasonic Inspection
    — Eddy Current Inspection

6. Welding procedure and operator qualification
7. Inspection of welding electrodes and filler metal

Chemical Analysis

The simplest and perhaps the most important quality control method is the analysis of the materials specified to verify that the material ordered is within the prescribed composition limits. This can be accomplished by specifying that a specimen representing the material ordered be analyzed at a company or commercial laboratory. *Mill test reports* and other documents from the manufacturer, such as certified test reports, can also be relied upon by the designer.

Leak Testing

Leak testing is accomplished mainly to determine soundness of equipment rather than to determine potential corrosion sites; however, leak testing can often locate such sites. Hydrostatic testing is probably the oldest testing method and is generally the most used.

Tanks are filled with water and after a predetermined time, usually 24 hours, they are inspected for any evidence of leakage. Where leaks are found, welding repairs are generally made. Precautions regarding the water used for testing is covered in Chapter 3.7.

Pressurized hydrostatic testing is also used. After the water has completely filled the tanks or enclosed vessels, pressure is increased to 1-1/2 times the expected working pressure and is held for a length of time, usually 1 hour. The outside surface of the equipment is then inspected for leaks.

When equipment is to be pressurized in service with a gas, pressurized air is sometimes used for testing at 1-1/2 times the working pressure. This is generally a hazardous procedure and should be conducted only under strict safety precautions.

Metallographic Examination

When materials of construction are to have a specific internal structure for the greatest corrosion resistance through heat treatment, surface treatment, working, etc., metallographic inspections should be made. A sample with the same composition as the material of construction can be treated through all the steps that the material of construction goes through. Sometimes, when practical, tabs are cut from the material itself for the same purpose. Such procedures must be clearly defined in specifications and in purchase orders.

Acceptance Testing

The following sections describe different tests that can be used to determine if materials of construction are acceptable or not:

Huey test (or the boiling nitric acid test). As discussed before, austenitic stainless steels, particularly low carbon stainless steel such as AISI 304L, are not only sensitive to variation in chemical composition but are also sensitive to the heat treatment and processing received at the mill. One way for the designer to determine if this has been properly done is for him to specify that a representative specimen of the heat of steel involved be given the so-called "boiling nitric acid test" specified in ASTM Standard A-262, Practice C. This test subjects stainless steel specimens for 48 hours in boiling 65% nitric acid.

When specifying this test, the designer should ensure that the specimen really represents the heat of steel involved. He should specify that a specimen be cut in the presence of a company inspector. A buyer usually has very few failures of austenitic stainless steel heats obtained directly from the steel mills, since it is the mills' custom to conduct their own boiling nitric acid tests before releasing the heats. The following example shows what can happen if a careful specimen-taking policy is not followed:

Example — A critical piece of equipment was to have been made of AISI 304L stainless steel. The heats of steel used in this equipment had all duly passed the boiling nitric acid test. For some reason, an inspector checked the individual plates that made up the equipment for molybdenum content with a spot tester after the process equipment had been completely fabricated. It turned out that more than half of the plates involved were actually AISI 316 stainless steel, which were unsuitable for the intended nitric acid service. Therefore, the equipment had to be scrapped. Because of the nickel shortage at that time, purchasing agents had been desperately buying stainless steel from any place that had it in stock. The place where the steel in question had been procured was visited; it was a small steel reseller in New York. We asked where the acceptance specimen had come from and the reseller pointed to a 5 x 5-ft (2 x 2-m) piece of stainless steel in the corner and said "I always cut the specimens from there." The FBI thought sabotage was involved. It was not. This reseller did not understand what the specimen was for, but he did know that to stay out of trouble, he should cut out specimens from the piece of "good" steel.

Strauss test for stainless steel. This test is not used as frequently as the Huey Test for testing lots of austenitic stainless steel. In this test, the specimen is exposed to boiling 3% cupric sulfate and 10% sulfuric acid for 72 hours to determine the susceptibility of heats of stainless steel to intergranular corrosion.

Test for heat-treatable aluminum alloys. Aluminum alloy specimens in the heat-treated condition are exposed to a solution of 1% hydrogen peroxide and 5% sodium chloride for 6 hours at 86 F (30 C) to determine susceptibility to intergranular corrosion.

Tests to Determine Crevices or Cracks

As discussed previously, crevices or cracks can cause accelerated local corrosion. These may be in the form of cracks, laps, shrinkage areas, seams, lack of bond, laminations, etc., that may be difficult or impossible to see with the naked eye. There are several tests capable of identifying such cracks or discontinuities that a designer can specify, such as:

Liquid penetrant inspection. This method is an easy, inexpensive method of detecting cracks, crevices, etc. (See Figure 6.1.) This method involves the following actions:

1. A red dye (the penetrant) is sprayed over the cleaned test area and migrates into any cavity, crack, or other discontinuity that is open to the surface of the metal.
2. The excess dye on the surface is removed, leaving only the red dye in the discontinuity.
3. A white "developer" is then sprayed on the surface.
4. The white developer blots or wicks up the red dye diffusing it so that the red dye indication is much wider than the crevice or discontinuity underneath. In this way, crevices, etc. are more easily detected. Figure 6.2 shows a shrinkage crack around the toe of a fillet weld that was not detected until the dye penetrant test was applied.

The designer can specify that the standard to be followed for this type of inspection is ASTM Standard E-165 ("Liquid Penetrant Inspection"). There are other standards on this subject that may also be used.

Magnetic particle inspection. This inspection method can detect small and shallow cracks in ferromagnetic materials. The principle of this test is that when the part or area under test is magnetized, cracks, discontinuities, etc. that lie in a direction generally transverse to the direction of the magnetic field will cause a *leakage field* to be

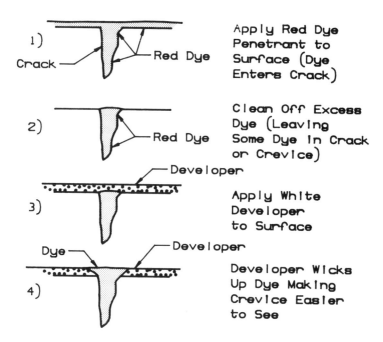

FIGURE 6.1 — *Dye penetrant technique; note the sketch above shows a cross section of a crack (or crevice) magnified about 20 to 40X.*

established. When finely divided steel particles are spread evenly over the test area, the particles over the cracks will orient themselves in a different direction from other particles in other noncracked areas, thus indicating precisely where the cracks are located.

This inspection method is only good on materials that can be magnetized. Non-ferromagnetic materials, such as aluminum, copper, copper alloys, titanium, titanium alloys, and all 300 Series or austenitic stainless steels, cannot be tested by this method. The designer can specify that parts to be inspected by this method should be in accordance with the American Society for Nondestructive Testing Recommended Practice SNT-TCla ("Magnetic Particle Inspection"). Again, there are other standards that may be used, such as ASTM Standards A-275 and A-456 ("Magnetic Particle Testing of Forgings"), and E109 ("Dry Powder Magnetic Particle Inspection").

Radiographic inspection. Radiographic inspection is an ex-

FIGURE 6.2 — *Shrinkage crack disclosed around the toe of a fillet weld by the dye penetrant test.*

tremely useful tool for detecting cracks, crevices, discontinuities, etc. in equipment, which can lead to localized corrosion. Because of its widespread use, this inspection method is emphasized here. The radiographic inspection also discloses subsurface flows, such as mechanical defects and unsound welds, which are perhaps its major inspection use. Radiographic inspection has proven especially useful for corrosion site detection at or near welds in the inside diameters of piping, tubing, and equipment where the liquid penetrant and magnetic particle inspections usually cannot be used. Much has been written about this rather complicated, and in some cases, hazardous procedure. For the sake of brevity, this text will be confined to the radiographic inspection of welded pipe. The designer is referred to a complete treatise on this subject in the *ASM Metals Handbook*, Vol. 11, pages 105 to 156 for further information on other applications similar to the pipe example. Also, refer to Section VIII of the ASME Pressure Vessel Code, which provides the most widely used rules covering the technique for radiographic examination of welded joints and rules for interpretation of radiographs.

Basically, radiographic inspection consists of directing either x-rays or gamma rays of such strength that they penetrate the pipe wall and welded areas and fall upon a sensitive photographic film. The amount of x-ray or gamma-ray penetration varies inversely with the thickness of the metal. Thinner sections, cracks, crevices, or voids allow the most radiation to pass, thus causing the film to be darker in those areas. The thicker the section, the less the amount of radiation passed and, hence, the film is much lighter in these areas. Cracks, crevices, etc. can be located and evaluated by observing or "reading" the film.

The radiation sources are either x-ray machines or radioactive materials. X-rays are produced when electrons traveling at a high speed collide with matter. In the usual type of x-ray tube, an incandescent filament supplies the electrons and thus forms the cathode, or negative electrode of the tube. The high voltage applied to the tube drives the electrons to the anode or target. The sudden stopping of these rapidly moving electrons in the surface of the target results in the generation of x-rays.

Gamma rays emanate from either natural radioactive material, such as radium or radioactive isotopes like Cesium 137, Iridium 192, and Cobalt 60. (Radium has not been used commercially for inspection purposes for many years.) There are other sources, but the last two isotopes are the most commonly used.

Figures 6.3 and 6.4[56] show how the source, either gamma ray or x-ray, is positioned for the inspections of large, medium, and small pipes and tubings. This technique can be used for either longitudinal welds or circumferential welds.

Figure 6.3 exhibits:

1. Film inside, radiation source outside is used for medium-size pipe.
2. Film outside, radiation source inside is used for large pipe. (A full circumference weld can be inspected at one time by placing film completely around the pipe.)
3. Film and radiation source outside is used primarily for small-diameter pipe. This figure shows a longitudinal weld being inspected.

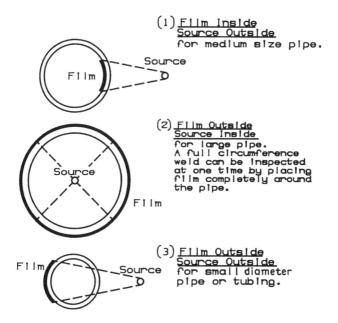

FIGURE 6.3 — *Gamma ray inspection of pipe.*

Figure 6.4 shows:

4. X-ray radiographing of multiple lengths of tubing or small diameter piping is an inexpensive way of examining many tubes at once.
5. When x-ray radiographing of medium diameter and small piping circumferential welds, the x-ray machine head is positioned at an angle so that the radiation will pass through the weld at an angle. In this way, only two exposures 90 degrees apart are required to produce an image representing the entire weld.
6. X-ray radiographing of medium and small longitudinally welded pipe.

Figure 6.5 shows how various defects are detected in radiographic examinations. The radiographs depicted are actual pictures of radiographs held up to the light.

Possible defects depicted in welds include:

Incomplete penetration —One of the most common defects found in welds is incomplete penetration. [See Figure 6.5(A).] In

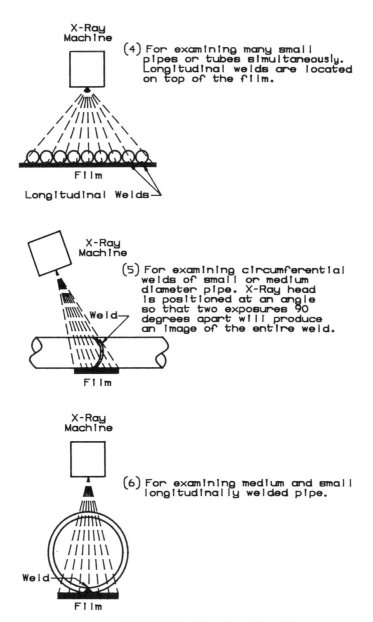

FIGURE 6.4 — X-ray inspection of pipe.

susceptible metals, localized accelerated corrosion could occur here. In Figure 6.5(A), Radiograph Nos. 1 and 2 show a lack of penetration within a short distance.

Undercuts — Undercuts on the face of a weld can usually be detected by the naked eye. However, they are often disclosed by radiographic inspection. In Figure 6.5(B), Radiograph Nos. 3 and 4 show weld undercut areas where accelerated corrosion could occur.

Shrinkage cracks —Shrinkage cracks appearing alone or in conjunction with porosity should be avoided for maximum corrosion resistance. In Figure 6.5(C), Radiograph No. 5 shows cracks emanating from porosity, and Radiograph No. 6 shows a long, shrinkage crack.

Porosity —Porosity, where it is open to the surface of the weld, can cause accelerated corrosion, as shown in Figure 6.5(D), Radiograph Nos. 7 and 8.

The radiographing procedure used in obtaining these radiographs is shown in Figure 6.4, Procedure No. 6.

Ultrasonic inspection. Ultrasonic testing is a nondestructive method for detecting defects in metals. Instruments are of two kinds based on pulse echo and resonance principles. The pulse echo type will be discussed here.

The pulse echo type generates high frequency (0.5 to 12.0 megacycles per second) sound waves electronically and transmits them through a nonmetallic piezo-electric crystal (such as quartz) mounted in a search unit, which is coupled with the part being tested. The search unit can be aimed either straight into, at an angle to, or at the surface of a part. The couplant (carries the sound from the search unit to the material being tested) can be water, light oil, or glycerine. The crystal vibrates while sending out ultrasonic pulses and then receives the reflection of those pulses from a defect or the back side of the part under test. The portion of the original pulse that is reflected back to the crystal is converted into electrical energy, which is amplified and transposed on the screen of a cathode ray tube. This amplified reflected wave appears as a vertical modification to the right of the indication set up by the initial impulse.

For a given ultrasonic velocity in a metal, a sweep across a part by a search unit can detect the location and size of various flaws. For critical parts, the designer can specify that this inspection method be applied at the mill where the materials of construction are being processed. Many, if not all, steel mills are using ultrasonic methods for

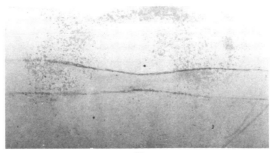

#1

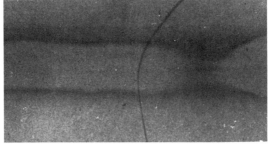

#2

View of Radiographs (Pictures of films taken in front of a source of light)

A. <u>INCOMPLETE PENETRATION</u>

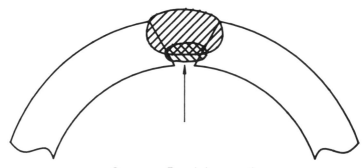

Cross Section of
Longitudinally Welded Pipe

FIGURE 6.5A — *Incomplete weld penetration.*

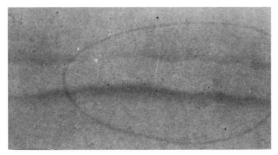

#3

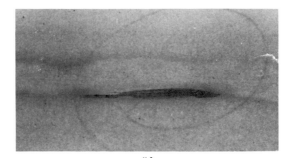

#4

View of Radiographs

B. <u>UNDERCUTS</u>

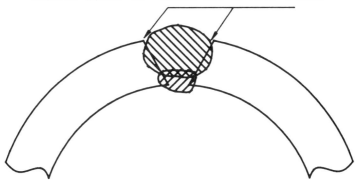

FIGURE 6.5B — *Undercuts.*

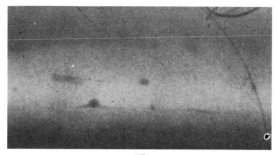
#5

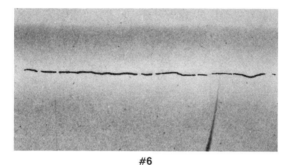
#6

View of Radiographs

C. <u>CRACKS</u>

CRACKS WILL APPEAR AS DARK LINES - SOMETIMES ASSOCIATED WITH POROSITY. ANY POROSITY WITH A TAIL IS CONSIDERED A CRACK.

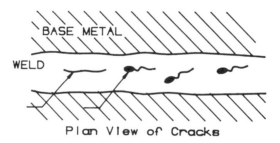

FIGURE 6.5C — *Cracks in welds*

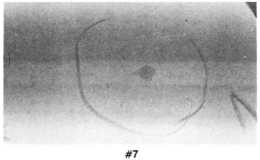

#7

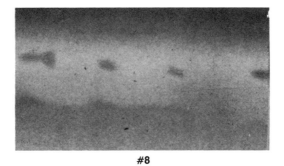

#8

View of Radiographs

D. <u>POROSITY</u>

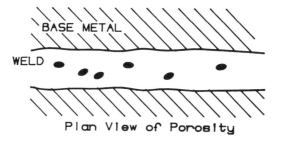

FIGURE 6.5D — *Porosity in welds.*

control of quality in rolled and forged carbon and alloy steel bar stock, forging billets, tubing, piping, and rolled plates. Plate that has been cladded with another material, such as stainless steel clad to carbon steel, can also be ultrasonic inspected to determine any areas of loose cladding.

There are many specifications governing ultrasonic inspection that the designer can use in his purchase orders. Military Standard MIL-STD-271B (Ships) defines procedures as well as acceptance limits. It is used widely throughout industry as an ultrasonic testing specification for use in purchasing mill items from vendors.

There are also a number of ASTM specifications dealing with ultrasonic testing. The acceptance-rejection limits are usually subject to agreement between the purchaser and the vendor in each specification, as follows:

>ASTM E164 — Method For Ultrasonic Contact Inspection of Weldments
>
>ASTM A-435 — Method and Specification for Ultrasonic Testing and Inspection of Steel Plates of Firebox and Higher Quality
>
>ASTM A-368 — Recommended Practices for Ultrasonic Testing and Inspection of Heavy Steel Forgings

Eddy Current Testing

Eddy current testing is used most in the detection of discontinuities in tubular products, such as heat exchanger tubing. As a quality control measure in the manufacture of metal tubing, it is common practice to pass such tubing through encircling coils of an eddy current tester.

In this type of inspection method, eddy currents are caused to flow in the tubing and are observed and compared with eddy currents that flow in a standard pipe which is considered acceptable. The test is complicated by variations in test or test sample conditions. However, when used by experienced tube mills, it can detect inside and outside diameter defects at speeds of over 100 ft (30 m)/min.

The defects that can be determined are inclusions, laps, splits, seams, gouges, dents, voids, cracks, tears, etc. Most tube mills that manufacture welded stainless steel tubing possess eddy current facilities for testing full-finished tubing on a production line basis. The method is not universally used in the seamless tube industry because ultrasonic testing seems to be preferred.

The size range for inspection by the eddy current method ranges from the smallest tubing sizes to about 3-in. (76-mm)-outside diameter with walls 1/2-in. (13-mm) thick.[57] ASTM Standard A-249 ("Welded Austenitic Steel Boiler, Superheater, Heat-Exchanger and Condenser Tubes") permits, when accepted by the purchaser, the use of the eddy current test as an alternate to the hydrostatic test. ASTM Standard A-450 ("General Requirements for Carbon, Ferritic Alloy, and Austenitic Alloy Steel Tubes") specifies that a nondestructive tester be used by the mill to detect any defects as specified. Such tests may be made on the welded seam and adjacent area or on the entire cross section of the tubing at the option of the producer.

Welding Procedure and Operator Qualification

The designer's original specifications reflected in the inspection briefs should specify a welding procedure and operator qualification tests. (An inspection brief is a document usually compiled by an inspection department based on a designer's specifications.) It is very important, of course, that welds be sound to enable maximum corrosion resistance. These qualification tests help assure acceptable welding. The designer should include the following statement in his specification, "All welding procedures and operators shall be qualified in accordance with the provisions set forth in the ASME Boiler Code Section IX or AWS specifications."

Inspection of Welding Electrodes and Filler Metal

Many corrosion failures have been caused by the mixup of electrodes or filler rods in the fabricators' bins. The manufacturer of welding electrodes and filler rods is required to make many tests prescribed by the AWS on his product before it is sealed into cartons. Consequently, the integrity of the welding electrodes can be relied upon but not after the cartons have been opened. For this reason, the designer should not only specify the types of electrodes required, but also that the company's inspector allow only unopened cartons of the electrodes to be used on the job and then those electrodes must be kept not only isolated from other welding jobs, but carefully marked so no mixups can occur.

Example of Excellent Quality Control

Figure 6.6 shows four waste storage tanks in various stages of erection at the Savannah River Plant. When completed, they were 85 ft, 0 in. (26 m) in diameter and 33 ft, 0 in. (10 m) high. Probably no storage tanks in history have experienced better quality control and, consequently, are excellent examples. Some of the major quality control steps, but not all, are described below:

FIGURE 6.6 — *Four waste storage tanks being built at the Savannah River Plant.*

1. Each heat of steel is verified as to chemical composition, mechanical properties, and steel mill procedure. This involved the entire fabrication history of each heat delineating the major fabricating steps, heat treatment, and microstructures.

2. Some of the heats of steel are drop-weight tested in accordance with ASTM Standard E-208 ("Conducting Drop-Weight Tests to Determine Nil-Ductility Transition Temperature in Ferritic Steels") to verify ductile fracture characteristics.

3. Each individual plate of steel is given an ultrasonic inspection test on a close-grid pattern to verify soundness.

4. The surface of each plate is visually inspected for conformity to specified minimum surface defects. Mill scale is verified to have been removed.

5. All welding procedures and welding operators are qualified in accordance with the ASME Boiler Pressure Vessel Code Section IX requirements. This includes automatic welding, as well as manual welding.

6. Welds are given 99.9% radiographic inspection. Those welds difficult or impossible to radiograph because of the geometry or weld type are vacuum tested. (Soap suds are applied on a local area and are vacuum drawn. Any bubbles or bubbling indicates an unsound weld.)
7. As welding proceeds, the welds are radiographed and the results are noted on a master sheet that shows all the welds in the tank. If a weld is found to be defective, it is ground out and radiographed again. This continues until each weld is found to be acceptable. All this information goes on the master sheet along with the heat history and the numbers of individual plates. The individual welding operators who performed the work are also identified on the sheet for each weld listed.
8. The radiographic inspections and evaluations are performed by an outside laboratory not connected with either the tank fabricator or the plant.
9. All tanks are stress relieved at a minimum of 1100 F (598 C) for 1 hour. During the stress-relieving heat treatment, many thermocouples throughout the tank monitor the progress of the operation. Strict limits are set as to the maximum allowable variation in temperature from point to point in the tank. Strict limits are also set as to the maximum allowable heating rate above 600 F (315 C) to the stress-relieving temperature and the maximum allowable cooling rate from the stress-relieving temperature to 600 F (315 C). (See Chapter 4.2.)
10. After the heat treatment has been completed and the tank has been cleaned, the steel surfaces are inspected for compliance with cleanliness specifications.
11. All radiographs and pertinent documents for a specific tank are stored in one trunk. Thus, pertinent fabricating and inspection data are always readily available.

Heat Treatment Verification

Because the heat treatment of metals and alloys often affect corrosion resistance, it is essential that the designer impose some manner of quality control on heat treatment operations. The importance of assuring that the proper austenitizing temperature and time at heat, the type of quenching media, the temperature, the tempering temperature, and the time at heat maintained cannot be overemphasized.

However, it is also essential to assure that the heat-treating equipment is in good operating condition. For instance, standard temperature thermocouples can be used to determine if a furnace is actually operating at the set temperature. Furnaces have been found by this procedure to be operating at temperatures hundreds of degrees off the designated set temperature. Competent heat treaters routinely check their furnace temperatures and therefore, their records may be used by the inspectors as verification.

On critical jobs, the designer can specify that specimens of the same material involved are heat treated along with the actual process equipment or part. In this way, the specimen after heat treatment may be sectioned, polished, etched, and observed under the microscope to verify that the required microstructure has been obtained. When appropriate, the hardness of the process equipment or part may be determined and compared with the specified hardness.

☐ 6.2 Corrosion Monitoring

When process equipment has been properly designed, built, inspected, and installed in a plant, the job of insuring maximum corrosion resistance is not necessarily complete. Subsequent changes in process solutions, such as variations in pH, temperature, velocity, reagents, inhibitors, etc., during operation may drastically alter the corrosion resistance of the materials of construction. In many cases where corrosion testing of candidate materials was not precise in the first place (perhaps because of the difficulty of duplicating the actual process conditions), the reliability of the materials of construction may be questionable.

Usually, good process management requires the analysis of process streams and a review of operating data. This can possibly determine any abnormalities in the corrosion resistance of materials of construction; however, such procedures have definite limitations for corrosion monitoring, and better corrosion control may be needed. For this reason, a corrosion monitoring program is often specified during the design stage. Such a program is especially necessary for new processes or new equipment in critical corrosive conditions where leaks of toxic nature are completely unacceptable or if irregular or unusual conditions are anticipated. The designer should assure that these corrosion monitoring programs are thoroughly outlined in the specifications and purchase orders. The most popular method of measuring the effects of corrosion in plant streams is to expose corrosion coupons at strategic locations.

Corrosion Specimen Monitoring

The corrosion coupons discussed in Chapter 2 for laboratory and pilot plant use can be used for monitoring corrosion in plants, except that multiple coupons are usually exposed so that coupons of the same materials of construction can be removed one at a time over periods of time to determine any changes in corrosion behavior. Figures 6.7 and 6.8 display corrosion racks for corrosion monitoring use. Figure 6.9 shows two corrosion racks designed for very limited space.

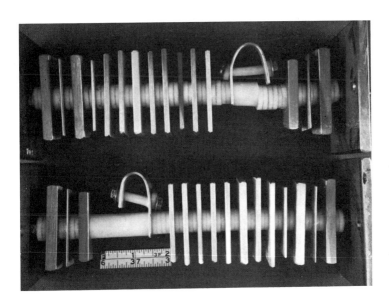

FIGURE 6.7 — *Two corrosion racks used for plant monitoring.*

The disadvantages of using coupons to monitor plant corrosion conditions are as follows:

1. At least several days of exposure are required to obtain corrosion rates. At best, this method gives average results, meaning that sudden and short-lived high corrosion incidents (resulting from incorrect process additions, loss of inhibitor strength, etc.) may be completely missed.

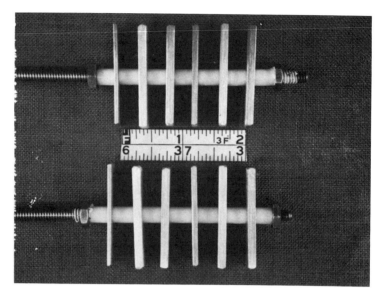

FIGURE 6.8 — Smaller corrosion racks.

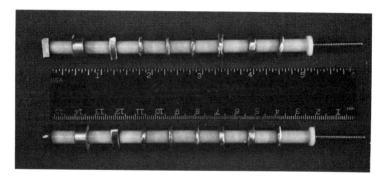

FIGURE 6.9 — Corrosion racks for limited space.

2. Also, in many processes, shutdowns are required to retrieve the corrosion coupons. This can be expensive and time consuming.
3. In addition, coupons have to be carefully prepared before installation and after exposure. They must be cleaned, dried, weighed, and measured, and corrosion rates must be calculated. This all takes time.

Equipment Inspection

Another method of corrosion monitoring is to make thickness tests and surface observations on actual process equipment. The thickness tests can be made using radiation or ultrasonic measuring devices. This type of corrosion monitoring, although the most direct method, requires periodic equipment shutdowns and decontamination (when needed), and presents access problems. Trained personnel are required for these observations, since the results are highly dependent upon the inspector's interpretation.

Considering the disadvantages and limitations of corrosion monitoring, there has been a basic industrial need for methods to continuously measure corrosion directly in the process stream.

Instrument Monitoring

Modern electrical and electrochemical methods of corrosion monitoring have been developed and commercialized to provide tools for rapid testing under plant and field conditions. These methods are unique because they can monitor the corrosion remotely and continuously without requiring plant shutdowns. Because monitoring can be continuous, short-term detection of changes in corrosion rates, even minute ones, can be identified with specific plant procedures, additions, or upsets. The following two types of corrosion monitoring equipment are available for industrial use:[58]

Electrical resistance (ER) type. Measuring corrosion by electrical resistance (ER) is based on the change of electrical resistance of a metallic specimen as its cross section decreases from loss of metal resulting from corrosion. The electrical resistance measure is not related to the chemistry of the corrosion reaction in any way. The measurement is simply based on the diameter of the sensor.

The electrical resistance method was developed in the late 1950s by a group of American Oil and Pure Oil Corrosion Engineers and has since been commercialized by several companies. In general, the ER system includes a probe made up of a sensor composed of the alloy to be tested, a connecting cable, and an instrument package that may include a recorder. Mathematic differentiation is required to determine corrosion rates.

The sensors or probes can be in the form of loops of wire, tubing, or strips, depending upon the sensitivity desired. The probes are made of the metal or alloy to be corrosion evaluated. Probe designs have been developed to permit operation up to 1000 F (538 C) and pressures to

5000 pounds per square inch (34,475,000 Pascal). Figure 6.10(a) shows a loop ER probe. Figure 6.10(b) exhibits the protective sheath for the loop. In operation, the loop is installed within the sheath. Figure 6.11 displays an ER probe that is situated inside a welded tube. This is the most commonly used type of probe.

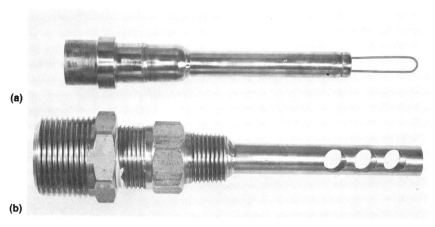

FIGURE 6.10 — *Electrical resistance loop probe; (a) loop probe and (b) protective sheath.*

FIGURE 6.11 — *ER probe of the welded tubing type.*

When the designer wishes to specify the ER type of corrosion monitoring equipment, he should first discuss this with a representative of the producer of such equipment. By doing this, the practicality of the use of ER equipment and the availability of appropriate sensors, etc. can be determined. Rohrback Cosasco Systems (Santa Fe Springs, California) manufactures an ER type of corrosion monitoring equipment called a *Corrosometer*, which is probably the most widely used equipment of this type. The Petrolite Equipment and Instrument Group (Houston, Texas) also produces quality ER equipment.

Linear polarization rate (LPR) type. As discussed in Chapter 1, corrosion of aqueous solutions is electrochemical in nature, differing from ordinary chemical reactions in that there is a flow of electricity through a finite portion of the corroding metal. This phenomena is used to determine corrosion rates. This method is called the linear polarization rate (LPR) method (sometimes called the *polarization resistance method* or the *instant corrosion rate method*). The amount of externally applied DC current required to change a freely corroding specimen's corrosion potential by about 10 millivolts is carefully measured. The amount of current is related to the corrosion rate. In other words, a DC current is applied from the testing equipment, which causes a specific amount of electrochemical polarization at the probes. The electrical flow associated with this polarization is proportional to the corrosion rate.

Instruments are manufactured with either two or three electrodes. The electrodes are made of the same metal to be tested. The readout in this type of corrosion monitoring device is in mils per year corrosion rate. Several companies manufacture LPR equipment. Rohrback Cosasco Systems sell their LPR instruments under the registered trade name of *Corraters*. The Petrolite Equipment and Instrument Group also markets LPR corrosion monitoring equipment. Figure 6.12 shows a Rohrback Cosasco Fixed LPR Probe. Figure 6.13 exhibits a retractable LPR probe, and Figure 6.14 displays a Petrolite Fixed three Probe LPR unit.

FIGURE 6.12 — *Rohrback Cosasco Systems fixed LPR probe.*

For meaningful measurements, the probes in each type of corrosion monitoring instrument must have the right composition of sensors. These sensors must be properly placed in the equipment, and the

FIGURE 6.13 — *Rohrback Cosasco Systems retractable LPR probe.*

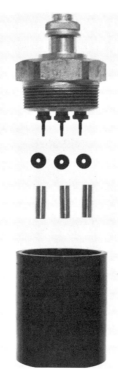

FIGURE 6.14 — *Petrolite fixed three-probe LPR unit.*

equipment must be calibrated accurately. For these reasons, manufacturers and experts in the field should be consulted by the designer before specifying specific equipment, probes, and probe locations.

Corrosion Monitoring Applications

Described below are various corrosion monitoring applications, including:

1. Extensive evaluation of the effectiveness and the persistence of various types of inhibitors. This includes inhibitors for pickling baths, automotive cooling systems, oil field solutions, and industrial plants.
2. Sensors have been installed in process streams to sound an alarm when steam leaks occur, which can cause excessive corrosion.
3. Successful determination of the effect of changes in processes on the existing materials of construction, thus providing confidence in making the change permanent without excessive corrosion.
4. Sensors installed in process streams have been used to diagnose the cause of abnormal corrosion caused by various contaminants in the streams.
5. Protection against over-neutralization of acids in process streams.
6. Performance evaluation of water treatment equipment before and after treatment.

6.3 Cathodic Protection

As discussed in Chapter 1.4, the action of a cell will be stopped if an outside direct electrical current is impressed on the anode, which overcomes the anodic corrosion current and a cathodic current results. This principle is extensively used in industry and is termed *cathodic protection*. Cathodic protection is defined as reducing or eliminating corrosion by forcing the metal (to be protected) to be the cathode via an impressed direct current or attachment to a sacrificial anode. Corroding structures have anodic and cathodic areas. When cathodic protection is applied to such structures, the whole structure becomes cathodic with no anodic areas. Hence, corrosion is stopped. In other words, cathodic protection is the passage of an electrical current into a metal at an equal or greater rate than a current would flow out of the metal if it were corroding.

When a metal is corroding, a direct current flows, which is limited by the resistivity of the electrolyte (the environment) and the degree of polarization at the anodic and cathodic areas. Corrosion occurs at the area where the direct current discharges *from* the metal *into* the environment. When the current flows *from* the environment *to* the metal, there is no corrosion. Therefore, the objective of cathodic protection is to force the entire metal structure to receive current from the environment. (To avoid confusion, remember the discussion in Chapter 1. The electrical current flows in the *opposite* direction from the electron flow.)

There are many excellent articles and texts dealing with this sophisticated protection method. Chapter 9, "Cathodic Protection," of NACE *Corrosion Basics — An Introduction* and NACE Recommended Practice RP0169-83 ("Control of External Corrosion on Underground or Submerged Metallic Piping Systems") are recommended. Because of the complexities of this method, the designer should consult experts when it appears that cathodic protection can be justified for a given application.

Cathodic Protection Application

The following passages are short descriptions of how cathodic protection is applied, used, and can be used by the designer. To help the designer visualize how cathodic protection is applied, three applications of buried structures are discussed, *unprotected buried tank, buried tank protected by an impressed current,* and *buried tank protected by sacrificial anodes.*

Unprotected buried tank. Figure 6.15 displays a buried tank that is corroding. Many corrosion cells have formed on the surface of the structure, accounting for many anodic and cathodic areas; however, for the sake of clarity, only one anode and one cathode is shown. The soil acts as the electrolyte carrying the corrosion current from the anode to the cathode. The buried structure itself acts as the current conduit to

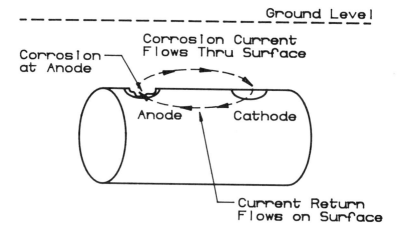

FIGURE 6.15 —*Example of buried tank, with no protection, corroding (simplified sketch showing one anode and one cathode—actually, thousands of such sites would exist).*

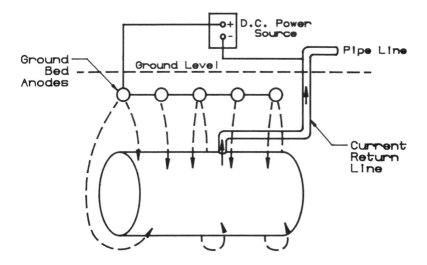

FIGURE 6.16 — *Example of a buried tank being protected from corrosion by an impressed current with ground bed anodes.*

return the current to the anode. Thus, the anodic areas are corroded, while the cathodic areas are protected.

Buried tank protected by an impressed current. Figure 6.16 displays a buried tank being protected by an electrical current supplied from a DC power source. The most common DC source is a rectifier. If AC power for the rectifier is not available, motor-generator sets (driven by gasoline engines) or batteries may be used. The positive terminal of the power source is connected to a group of anodes called *ground bed anodes* located underground in the vicinity of the buried structure. The negative terminal of the power source is connected by a wire to a pipe, which, in turn, is connected to the buried tank. When the current is turned on the ground bed, anodes are forced to generate a current into the ground and, hence, to the surfaces of the buried tank. The pipe (or in the case of buried pipelines, an insulated wire) acts as a return line conducting the current back to the power supply, thus completing the circuit. The result is that the entire surface of the buried structure has been made the cathode, and hence, no corrosion occurs because all the former anodes on the surface of the buried structure are converted to cathodes. The new anodes are now the ground bed anodes, which are corroded instead of the buried tank. Materials for the ground bed anodes can be steel, 14% silicon cast iron and carbon or graphite. The rate at which the anodes are corroded varies greatly with the material involved. For instance, cast iron has a rate roughly 1/50 that of steel, while graphite has a rate of about 1/25 that of steel. This type of protection is generally used when the current requirements and electrolyte resistivity are high.

Buried tank protected by a sacrificial galvanic anode. Figure 6.17 shows the tank being protected by a sacrificial galvanic anode. This is probably the most popular method of cathodic protection. The tank must be positive with respect to the galvanic anode selected before the anode can discharge a current. For example, if the buried structure is steel (which is usually the case), those materials above iron (or more active than iron) in the galvanic series are eligible materials for use as sacrificial anodes. This would include magnesium, aluminum, zinc, and their alloys. These are the materials generally used.

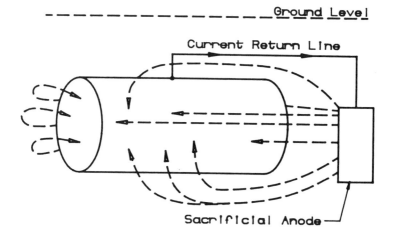

FIGURE 6.17 — *Example of a buried tank being protected from corrosion by a sacrificial anode.*

The sacrificial galvanic anode is buried near the buried structure. (Depending on the size of the buried structure, more than one anode may be required.) Again, the soil acts as the electrolyte conducting the current to the surface of the buried structure making it the cathode, as was the case with the impressed current. The buried tank is connected by an insulated wire to the sacrificial anode. Thus, the wire acts as the current return line. The sacrificial anode is thus corroded, and the tank is protected. This type of protection is generally used when current requirements are low and the electrolyte has relatively low resistivity (below around 8000 ohm-cm).

Structures submerged in water. Structures in a water environment may be protected by either impressed current systems or by sacrificial galvanic anodes, much like the application noted above. However, when sacrificial galvanic anode systems are used, much heavier anodes are usually used because high currents may be obtained as a result of the low resistivity of the seawater. The heavier anodes are also required to assure a reasonably long life before replacement is required.

Rings of sacrificial galvanic anodes are sometimes used around pipelines submerged in bays, rivers, or the sea. These rings are evenly spaced on the pipeline and are connected directly to the circumference of the pipe.

Sacrificial galvanic anodes for underwater use are generally made from either zinc or certain aluminum alloys. The interiors of large pipelines carrying seawater or other corrosives are usually lined with a suitable coating and protected with strip-type anodes. If such a steel pipeline is not coated, it may require large amounts of current for protection furnished by anodes penetrating the pipe walls at spaced intervals.

Tankers filled with salt water as ballast are very susceptible to corrosion. Galvanic sacrificial anodes are normally used for their protection.

Hulls of ships are usually protected by either zinc or aluminum galvanic anodes or impressed current. Whichever are used, the systems should be designed to handle a wide range of current output. The demand for current will vary with the condition of the water. For instance, rapidly moving water while the ship is underway demands far more current to keep the hull in a polarized condition than quiescent water at the dock.

Uses for Cathodic Protection

Cathodic protection is used in many applications, including steel piles, well casings, interior parts of structures such as pumps, condenser water boxes, pipelines, storage tanks, and structures buried in the ground, as just discussed, and many marine applications such as pier supports, offshore drilling platforms, hulls of ships, ship propellers, and steel bulkheads. This method was first successfully used by the oil and gas industries in protecting buried gas and oil pipelines, and in marine applications. The use of cathodic protection has spread since then to many other areas, such as electrical generating stations and manufacturing plants. Although the chemical industry uses cathodic protection extensively on buried pipelines and tanks, its use for chemical plant equipment has been somewhat limited. The problem is maintaining a protective current uniformly throughout the interior of process units. If the geometry is complex or irregular, the corrosion may be decreased in some areas but increased in other areas. (Anodic protection discussed in the next section does not have this problem, but it cannot be used on all metals.) However, vessels of uniform geometry have been successfully shielded from corrosion by cathodic protection.

The greatest use for cathodic protection is to lower or eliminate corrosion in steel structures. However, other metals may be protected, such as those susceptible to crevice corrosion like some stainless steels and Monel. Even aluminum can be protected, but care must be exercised that the protective potentials are not allowed to go too high. If this occurs, the environment can become so alkaline that the aluminum will experience significant corrosion attack.

Copper-base metals may be protected efficiently. For instance, ship propellers are almost universally protected from cavitation (see Chapter 3.4) via the cathodic protection method. Sacrificial anodes are attached to the hulls of ships near the propellers.

The outside surface of a pipeline may be guarded by cathodic protection; however, this does not protect the *inside* surface of the pipe. The reason for this is that the protecting current flows on the surface of the pipe and does not penetrate through the wall of the pipe.

How the Designer Should Proceed with Cathodic Protection

Alternatives. Although the principle of cathodic protection is simple, the actual application is not. For this reason, when the designer has found an application that might result in cost savings, he should consult with the company's materials or corrosion engineer, an outside consultant, or a supplier of cathodic protection equipment.

Suppose the designer is designing a carbon steel pipeline that is to be buried. What should he specify? The following are alternatives the designer has; he can specify:

　　　1. No protection;

2. Application of protective coatings;
3. Cathodic protection of the bare steel;
4. Cathodic protection of the coated steel;
5. A more resistant material of construction; or
6. Installation of pipeline above ground.

Cathodic protection survey. To determine which alternative is the best, a cathodic protection survey should be conducted to determine whether cathodic protection is feasible and what the costs will be for not only the installation, but for the cost of maintaining the system as well. The survey would determine the following:

1. Electrical resistivity of the soil. Resistivity measurements in ohm-centimeters would be made at locations where sacrificial anodes or ground bed anodes would be located. The number, size, and type of sacrificial anodes required would be determined as well as the required ground bed anodes size and number, and the type of DC power source needed.
2. Minimum voltage required for adequate protection.
3. Total amount of current required.
4. Availability of AC electrical power to operate DC rectifiers.
5. Presence of other buried pipelines or structures in the vicinity that might affect the contemplated system.
6. Range in temperature of environments should be determined. Resistivities of waters and soils usually decrease when the temperature increases. The freezing of water or soil can cause extremely high resistivity compared to the unfrozen condition. For this reason, ground beds should be installed below freezing levels of soil or water.
7. Presence of any other cathodic protection system that might affect the contemplated system.
8. Presence of stray electrical currents, such as street car electrical systems, electric railroads, generators, or other electric devices producing stray currents would be determined. Stray currents complicate cathodic protection systems.
9. Required life of the system.

Final determination. Based on the cathodic protection survey, the designer should consider the following factors:

1. If the soil has a very high resistivity, or the pipeline is to be temporary or its life expectancy is short, perhaps no protection is required. However, in many industrial cases, it is not good

practice to bury steel pipe without protection.
2. Apply a protective coating to the pipeline. Most buried carbon steel pipeline should have some kind of a protective coating. These coatings vary from coal tar enamels to elaborately coated and wrapped pipe. In between, there are many protective coating systems available. The wrappings can be sheet plastics, elastomers, or treated paper. Compared with cathodic protection, protective coatings are the least costly but are not as dependable.
3. Use cathodic protection on uncoated pipe. Generally and as stated above, this method is more expensive than coating.
4. Use *both* protective coatings and cathodic protection. In many applications, this may be the least expensive alternative. The major cost of cathodic protection is the power cost or the cost of replacing anodes. Coated pipelines require only a very small fraction of the current required for bare pipelines depending, of course, on the integrity of the coating. As the coating starts to deteriorate over the years, this arrangement generally can protect the system with only small raises in the required energy for a considerable amount of time.
5. Specify a more resistant material, such as concrete pipe or reinforced plastic pipe.
6. If costs will be very high, consider locating the pipeline above ground where it has the advantage of less corrosion and easy inspection.

How to Know if the System is Working

After the cathodic protection system has been installed, potential measurements are taken to determine whether sufficient cathodic protection is being achieved or not. The basis for this is that if a current is flowing, there has to be a difference in potential between the soil and the buried structure. This potential is based on the voltage drop across the buried structure and takes into account the resistance of the structure, the soil and polarization that may occur on the surface of the buried structure. The designer should specify the procedure that will tell if the system is working prior to the installation of cathodic protection. To explain further:[59]

1. If just enough current were collected by the cathodic areas to polarize them exactly to the open circuit potential of the anodic areas, corrosion would cease because there would no longer be a driving force to cause a corrosion current to flow. This situation

is not usually feasible in actual practice, so a net current flow onto the original anodic areas is required to be sure the equipment is actually being protected.

2. Measuring this excess potential with a voltmeter is not easy to accomplish. For instance, at the bottom of Figure 6.18 a buried tank where the potential of the tank to the earth is desired is shown. Connecting one side of the voltmeter to a pipe leading from the tank can easily be done; however, connecting the voltmeter to the earth is more difficult. Just driving a steel stake in the ground for the connection is not feasible because the steel to earth potential is not stable. A reference electrode must therefore be used to join the voltmeter to the earth in order to have a stable potential at the junction.

3. At the top of Figure 6.18, a commonly used reference electrode is shown. A voltage measurement made on the high-resistance voltmeter involves two parts:

   — The potential or voltage between the electrode itself and the earth through the porous plug contact with copper sulfate (a stable connection); and

   — The potential between the buried tank and the earth.

   The first potential is a constant value; the second potential is variable. The latter potential is what is required to determine changes in the tank (or any buried structure) to earth potential. Since the first potential is of constant value, the two potentials do not have to be separated and the total voltage reading may be used.

4. A structure to earth standard has been formulated to aid in determining whether full cathodic protection is being achieved or not. For steel, this value is −0.85 volts with respect to a copper sulfate electrode. The negative sign means that the buried structure is negative to the earth and that the cathodic protection current is flowing onto the originally anodic areas.

5. Another method of determining whether a structure is being properly cathodically protected is to wire weighed specimens to the structure via digging a hole and then burying the specimens. After a period of time, the coupons may be retrieved, weighed, and a corrosion rate may be determined.

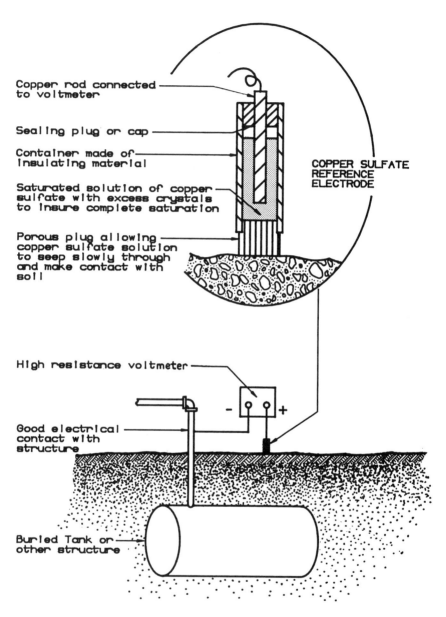

FIGURE 6.18 — *Measuring excess potential (voltage) to determine if the cathodic protection system is working properly. (Hook-up of cathodic protection system is not shown.)*

☐ 6.4 Anodic Protection

One problem with cathodic protection is its occasional inability to maintain a protective current uniformly throughout process units or equipment where the geometry is complex or irregular, as was mentioned in the last section. A fairly recent development (proposed in 1954) is the anodic protection method, which is the opposite of cathodic protection because an anodic current, rather than a cathodic current, is impressed on the equipment. Anodic protection does not have the disadvantage that cathodic protection has. In general, it can protect against corrosion in complex, irregular, or nonsymmetrical equipment. The reason for this is that in anodic protection, the *controlled* current flows from the anode through the electrolyte to the cathode. In *cathodic* protection, the opposite occurs; the current flows from the cathode to the anode. Because of the direction of current flow, anodic protection does not have the geometric limits found in cathodic protection. Surfaces many feet away from the cathode and sometimes narrow crevices can be anodically protected. This is the major advantage of anodic protection over cathodic protection.

When the designer is considering using anodic protection in his equipment or plant, a recently published book, *Anodic Protection: Theory and Practice in Prevention of Corrosion*[60] by Riggs and Locke, and edited by Hamner, will be helpful in making decisions.

How Anodic Protection Operates

First, the designer should realize that this protection method can be used only by *active-passive* metals and alloys. These metals include steels, stainless steels, aluminum, chromium, and titanium alloys. This protection method can only be used in certain environments since the composition of the electrolyte influences passivity.

This technique involves the application of an electrical current in a system susceptible to corrosion in such a way that the surface to be protected is made *anodic* to a cathode suspended in the corrosive solution. This application takes advantage of the Flade potential[61] or effect, which identifies the electrical condition of a surface that is passive because of a layer of electrons, oxides, or oxygen created by the electrical current. This protective layer can be maintained by the constant or intermittent application of small increments of current.

By contrast, a *cathodically* protected surface (discussed previously) is believed to be protected by a layer of hydroxyl ions or hydrogen resulting from electrical currents moving into the protected surface.

These layers in both cathodic and anodic protection systems passivate the surface with respect to the environment. As long as they are maintained by an electrical current, the passivity persists.

Identifying the Passive Zone

Problems with anodic protection are identifying those metals and alloys that exhibit active-passive characteristics in specific process solutions and then maintaining the protection in the passive area of those metal surfaces, not allowing drifts into active areas. As displayed in Figure 1.3, nonpassivating alloys cannot be protected by this method. On the contrary, such metals will experience accelerated corrosion under an *anodic* current.

Although the passive zone on which anodic protection is based has been known for many years, application of this principle in operating systems had to wait until the precision high current potentiostat described by Conger and Riggs,[62] which is capable of very precise control of potentials (voltages), was developed.

Laboratory tests. Laboratory tests using the precision high current potentiostat can be used to determine whether or not a combination of metals and corrosives will permit the use of anodic protection. By using a sample of the corrosive solution and a specimen of the materials of construction that are to be used in the equipment, a curve can be constructed from data generated by the potentiostat similar to that shown in Figure 6.19. This figure is a schematic of the electrochemical processes that occur during anodic polarization of an active-passive metal.

Anodic polarization curve generation. The ordinate of Figure 6.19 represents reference potentials, and the abscissa represents log of current densities. The anodic polarization curve can be constructed as follows. A sample of the materials in question is exposed to the test solution in a special laboratory setup that includes a potentiostat, a reference electrode (calomel), a vacuum tube voltmeter, an ammeter, and an auxiliary electrode (platinum). As the sample freely corrodes, its corrosion potential (E_{cor}) is established. The potential is then changed by using an external circuit to increase it in the oxidizing direction and then the generated current density is measured. This is displayed in Figure 6.19 as *Forward Scan* (the corrosion rate on the sample *increases*).The current generated is *proportional* to the corrosion rate; i.e., the greater the corrosion density, the greater the corrosion rate will be. The potential is increased until a *critical current density* (I_{crit}) is reached at a potential of E_p. At this point, polarization has been

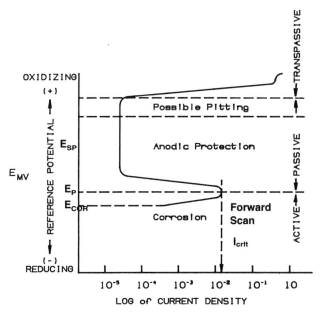

FIGURE 6.19 — *Potentiodynamic anodic polarization of an active-passive metal (from Reference 59).*

established, and the current density or corrosion rate *decreases* abruptly as the potential is increased. (This critical current density must be reached before active conditions are changed to passive conditions. The cost of anodic polarization is predicated to a large extent on this electrical requirement.) As the potential is slowly *increased*, the current density becomes *lower and lower* until it remains constant as the potential continues to be increased. Obviously, this is the vertical portion of the curve. At this region, the current density is lowered to about 1/300ths of the critical current density. This means that the corrosion rate has been drastically lowered. The potential in actual protection practice is usually maintained about half way up the vertical leg at E_{sp} (preferred). This gives some flexibility in both directions.

The shape of various polarization curves can vary greatly with the metal and corrodent involved.

(For clarification, in Chapter 1.4 on testing metal/environment conditions, oxidizing agents were added to increase the oxidizing power of a solution in order to explain passivity. In anodic protection, the increase in oxidizing power is occasioned by the increase in the electrical potential by an external circuit.)

In summary, the anodic polarization curve is typical of those that can be prepared, which not only establishes the possibility of using anodic protection, but also permits calculating the probable current consumption required to maintain passivity. The *passive zone* indicated in such curves depicts the condition of the surface which will resist the corrosive solution. As mentioned previously, surfaces maintained in an electrical condition within the limits of this zone will usually corrode at a rate significantly lower than that of an unprotected surface in some combinations of corrosive solutions and materials. However, the corrosion rate tends to increase as temperature, solution concentration, and turbulence increase, like a *cathodic protection* corrosion rate increases under these influences.

Anodic Protection Uses

Anodic protection has been used for more than 20 years to protect hundreds of tanks containing sulfuric acid, as well as for many heat exchangers used for acid production. It has been used to lessen corrosion rates for storing 93 to 99% sulfuric acid. The intent is to minimize iron pickup, as well as to extend service life. If proper passivation is reached, the corrosion rate can be lowered significantly compared to having no anodic protection.

Actually, iron in sulfuric acid lowers the aggressiveness of the acid on steel. As sulfuric acid with lower and lower iron content is produced, the need for protection of steel is increased.

Anodic protection can be used to widen the range of temperatures where stainless steel can be used in sulfuric acid. The range of sulfuric acid concentration that can be handled by stainless steel can also be increased. Approximately 400 anodically protected stainless steel coolers have been installed in acid recirculation loops in sulfuric acid plants throughout the world.[60]

Anodic protection is also used to protect carbon and chromium nickel steels against a wide range of other acidic and alkaline solutions. It has been used to protect mild steel in paper mill black acids, and spent alkylation acid, ammonia, ammonium nitrate with and without ammonium hydroxide, and other fertilizer solutions. It has also been applied to titanium, aluminum, chromium, and their alloys.

The Anodic Protection System

A typical anodic protection system is depicted in Figure 6.20. In certain systems, more than one cathode may be required. The cathodes are made of metals or oxides substantially inert in the solution and are automatically protected cathodically. The reference electrode is used as a base for comparison to determine the potential of the vessel wall. When this potential reaches a predetermined minimum, the control circuit is activated to cause the power source to add energy sufficient to bring the surface back to the proper passivation potential. Additional electrical energy can be applied in response to changes in the area of surface exposed to the corrosive solution, temperature, turbulence, and to some extent, to changes in solution chemistry as well as concentration.

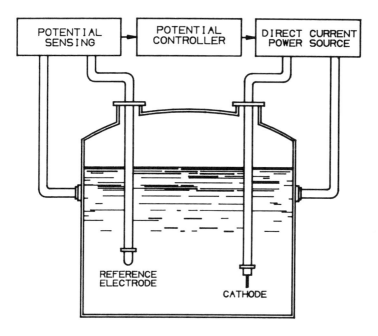

FIGURE 6.20 — *Anodic protection system schematic (from Reference 60).*

Figure 6.21 shows the corrosion rates of an actual carbon steel tank containing 93% sulfuric acid. The corrosion rates up the 48 ft (15 m) height of the tank wall show corrosion rates with and without anodic protection. In this example, the average corrosion rate of steel in 93% sulfuric acid is about four times the average corrosion rate of the steel when guarded by anodic protection.

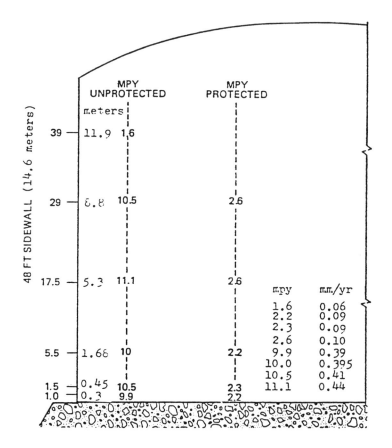

FIGURE 6.21 —*Effectiveness of anodic protection in a 93% sulfuric acid storage tank.*[60]

Passivation Persistence

Passivation of surfaces induced by anodic protection may be persistent after the current has been cut off, ranging in time from seconds in some highly corrosive systems to hours in a steel-ammonia fertilizer combination.

This persistence also permits using one set of control and power circuits to passivate more than one tank. In this mode, the sensing circuitry is switched from one reference electrode to another, either on a time schedule or on a demand basis so that current increments can be added to each tank surface from one power source. This divides the cost of control and power installations.

Economics of Installations

The cost of an anodic protection system itself can be determined and the cost of power established in advance with reasonable precision. When contrasted to alternative methods of corrosion control, the size factor becomes important because the cost of anodic protection does not increase directly as a function of size, whereas, the cost of coatings systems, for example, will.

How the Designer Should Proceed with Anodic Protection

The following are steps a designer should take when initiating anodic protection:

1. Have experts who can determine whether or not anodic protection will be effective for the contemplated service develop the anodic polarization curves.

2. When determining whether or not the conditions are active-passive with the proposed material of construction and the contemplated process, make sure that the solution and metals used for the test are actually representative. Remember, if several materials of construction are involved, such as stainless steel or Alloy 20 Cb-3 inside a steel tank that is anodically protected from sulfuric acid, the *passive* protection potential for the steel may overlap the *active* regions of these alloys.

3. Carefully compare anodic protection to cathodic protection and to other forms of protection as well. This comparison should be based on corrosion test results and should include the initial installation cost, the operational costs, the integrity of each protection system, and the value of money.

4. Consult outside consultants and/or equipment vendors to furnish pertinent information and recommendations.

6.5 Inhibitors

A corrosion control tool available to the designer is the corrosion inhibitor. *Corrosion inhibitors* are materials, both organic and inorganic, that are added to a water source or other fluids or gases in small amounts to reduce or stop corrosion.

The application of inhibitors must be viewed with caution by the designer, because like anodic protection, discussed before, inhibitors may afford excellent protection for one metal in a specific system but may aggravate corrosion for other metals in the same system. Since many water systems can contain several different metals and alloys, this fact should be remembered.

Inhibitor Classifications

Inhibitors have been classified in many ways, including by composition, mechanism of action, or form. There are also substances that retard corrosion by forming protective precipitates or by removing an aggressive constituent from the environment; these are also classified as inhibitors. To simplify a long list of classifications, the major classifications are listed below.

Organic inhibitors. Organic inhibitors affect the entire surface of the metal by adsorption on the metal, thus blocking both anodic and cathodic reactions. This type of inhibitor would include amines and sulfonates. Gelatin and glue are also examples of this category. Certain foods have built-in natural inhibitors; otherwise, cooking utensils might corrode at a greater rate than they would if no inhibitors were present. Organic inhibitors are commonly used in acid storage and metal pickling operations.

Vapor phase inhibitors. Vapor phase inhibitors carry inhibitors to metal surfaces by volatization. Cyclohexlamine nitrite and dicyclohexylamine carbonates are examples of this type of inhibitor. These inhibitors are used mainly in closed systems and in packaging. Packaging materials are impregnated with inhibitors such as sodium benzoate to reduce or eliminate corrosion because their high vapor pressures allows them to spread into the atmosphere and then be adsorbed on any metal surfaces. The inhibitor is applied in powder form or liquid.

Anodic inhibitors. Anodic inhibitors function by forming an insoluble protective film on anodic surfaces or by the mechanism of adsorption of the inhibitor on the metal. Thus, they passivate the metal. Examples of this kind of inhibitor are chromates, alkali phosphates, carbonates, silicates, and molybdates. Certain anodic inhibitors can cause accelerated corrosion and pitting attack if used in insufficient concentrations.

Cathodic inhibitors. Cathodic inhibitors function by forming an insoluble film on cathodic surfaces. These inhibitors are also adsorbed on the cathodic surfaces. Examples of cathodic inhibitors, which are generally less effective (but safer) than the anodic inhibitors, are zinc and salts of antimony, manganese, nickel, and magnesium.

Inhibitors based on oxygen removal. Boiler feed waters are treated with sodium sulfite or hydrazine to reduce or eliminate dissolved oxygen in the water. These scavengers are most effective in neutral or slightly acidic conditions. They are usually used in closed systems.

Additions of oxidizers. Oxidizers can be considered corrosion inhibitors under specific conditions. Some metals and alloys need oxygen for corrosion protection. Austenitic stainless steels (300 Series), Hastelloy C-276, titanium, and aluminum depend on a passive oxide film and are therefore aided by increased oxygen or oxidizing agents.

For example, *cone washers* are used for cleaning and brightening cartridge brass shell casings when making ammunition. These cone washers are made of cast, 18-8 stainless steel and handle 15% sulfuric acid at 150 F (66 C) and have been used for many years. It would appear that austenitic stainless steel is a very poor choice for this environment since sulfuric acid of this strength and temperature will attack the stainless steel, because the acid breaks down the protective film causing the stainless steel to become active. The reason this does not happen in the example is that copper dissolved from the cartridge cases acted as an oxidizing agent, keeping the oxide film on the stainless steel intact so that the stainless steel remains in a passive corrosion-resistant state.

How the Designer Can Use Corrosion Inhibitors

As can be seen, the designer is confronted with a great number of inhibitors and only a few of them have been discussed here. Table 6.1[62] summarizes various corrosive systems and the inhibitors that have been successfully used in industry to protect materials of construction. Specific systems are listed with the metals protected and the type of inhibitors used along with the proper concentration.

Inhibitors are generally specific in their actions, and the proper inhibitor choice must be made for each individual application. For this reason, outside consultants or specialists in the field of inhibitors should be used to help the designer meet his goals. To assist the consultant,

TABLE 6.1 — Various Corrosive Systems and the Inhibitors That Have Been Used to Protect Them[1]

System	Inhibitor	Metals Protected	Concentration	System	Inhibitor	Metals Protected	Concentration
Water, Potable	$Ca(HCO_3)_2$	Steel, Cast Iron + Others	10 ppm	Engine Coolants	Na_2CrO_4	Fe, Pb, Cu, Zn	0.1 to 1%
	Polyphosphate	Fe, Zn, Cu	5 to 10 ppm		$NaNO_2$	Fe	0.1 to 1%
					Borax	Fe	1%
	$Ca(OH)_2$	Al	Sufficient for pH 8.0	Glycol/Water	Borax + Mercaptobenzothiazole	All	1% + 0.1%
	Na_2SiO_3	Fe, Zn, Cu	10 to 20 ppm	Acids, HCl	Ethylaniline	Fe	0.5%
Water, Cooling	$Ca(HCO_3)_2$	Steel, Cast Iron + Others	10 ppm		Mercaptobenzothiazole	Fe	1%
	Na_2CrO_4	Fe, Zn, Cu	0.1%		Pyridine + Phenylhydrazine	Fe	0.5% + 0.5%
	$NaNO_2$	Fe	0.05%		Rosin Amine + Ethylene Oxide	Fe	0.2%
	NaH_2PO_4	Fe	1%	H_2SO_4	Phenylacridine	Fe	0.5%
	Morpholine	Fe	0.2%				
Boilers	NaH_2PO_4	Fe, Zn, Cu	10 ppm	Con. H_3PO_4 Most Acids	Thiourea	Fe	1%
	Polyphosphate	Fe, Zn, Cu	10 ppm		Sulfonated Castor Oil	Fe	0.5 to 1%
	Morpholine	Fe	Variable		As_2O_3	Fe	0.5%
	Hydrazine	Fe	O_2 Scavenger		Na_3AsO_4	Fe	0.5%
	Ammonia	Fe	Neutralizer				
	Octadecylamine	Fe	Variable				
Brines	$Ca(HCO_3)_2$	Fe, Cu, Zn	10 ppm	Vapor Condensate	Morpholine	Fe	Variable
	Na_2CrO_4	Fe, Cu, Zn	0.1%		Ammonia	Fe	Variable
	Sodium Benzoate	Fe	0.5%		Ethylenediamine	Fe	Variable
	$NaNO_2$	Fe	(NaCl, 5%)		Cyclohexylamine	Fe	Variable
Oil Field Brines	Na_2SiO_4	Fe	0.01%	Enclosed Atmosphere	Cyclohexylamine Carbonate	Fe	1 lb per 500 cu ft
	Na_2SO_3 (or SO_2)	Fe	O_2 Scavenger (O_2 x 9) ppm		Dicyclohexylamine Nitrite	Fe	1 lb per 500 sq ft
	Quarternaries	Fe	10 to 25 ppm		Amylamine Benzoate	Fe	Variable
	Imidazoline	Fe	10 to 25 ppm		Diisopropylamine Nitrite	Fe	Variable
	Rosin Amine Acetate	Fe	5 to 25 ppm		Methylcyclohexylamine Carbonate	Fe	Variable
	Coco Amine Acetate	Fe	5 to 15 ppm				
	Formaldehyde	Fe	50 to 100 ppm				
Seawater	Na_2SiO_3	Zn	10 ppm	Coating Inhibitors	$ZnCrO_4$ (yellow)	Fe, Zn, Cu	Variable
	$Ca(HCO_3)_2$	All	pH Dependent		$CaCrO_4$ (white)	Fe, Zn, Cu	Variable
	$NaH_2PO_4 + NaNO_2$	Fe	10 ppm + 0.3%		Red Lead	Fe	Variable
	Na_2NO_2	Fe	0.5%				

[1]From Reference 61.

the designer can do a number of things to prepare for the consultant's visit. The following items apply to water inhibition; however, the approach would generally be the same for steam, acids, gases, and other fluids:

Review all water systems in your design. Determine how aggressive each of the water systems are based on past experience, laboratory tests, or pilot plant tests. (Based on such information and knowing the composition, temperature, velocity, coupled with knowledge of the materials of construction involved, a judgment can be made with the consultant as to whether or not inhibitors are required or are feasible to use.)

Decide between once-through and closed recirculating water systems. Decide whether to design the water systems as once-through or closed recirculating systems. For the following three reasons, the once-through or open systems are not used extensively today:

— Obviously, such systems use far more water than the closed system.
— Certain inhibitors, such as sodium dichromate and other chromates, which are valuable inhibitors, are prohibited by state and federal antipollution laws from being discharged from plants.
— Lastly, because of the large amount of inhibitor required, the cost of inhibitors in open systems might prove to be prohibitive.

Determine the water composition. Of particular importance is the pH level and the amount of dissolved solids present. If the pH level is on the acid side, soda ash, ammonia, lime, alkaline phosphates, and other materials may be required to neutralize the acidity so that the inhibitor will be most effective. If the dissolved solids are high, the conductivity of the water will be high, which means it will have greater capability of carrying a corrosion current. Under such conditions, the formation of a protective film by the inhibitor will be interfered with and the corrosion rate can rise.

Determine the water temperature and velocity. The higher the temperature is, the greater the corrosion rate may be and this may limit the effectiveness of a selected inhibitor. On the other hand, increased velocity (within reasonable limits) improves inhibitor effectiveness.

Study the materials of construction selected for the water system. Make a list of all the materials, both nonmetallic and metallic. The best system, when practical, has only one material of construction. The worst system has many dissimilar metals and galvanic couples.

Remember that passivation of one metal in a couple by the inhibitor can accelerate corrosion in the other metal by changing the cathode to the anode ratio. Keep in mind that corrosion in crevices is difficult to control by inhibitors.

Determine the cleanliness of the system involved. Dirty surfaces can cause inhibitors to be ineffective and can also cause clogging by reaction with the inhibitor. Insist that the system be cleaned periodically if dirt is likely to accumulate.

Review the use of the water. Process use, boiler feed, cooling towers, circulating systems for cooling or air conditioning and steam condensate are the most common water uses. Each has its own corrosion problems which must be considered.

After obtaining the consultant's recommendations, the designer should take the following actions:

1. **Contact inhibitor suppliers for recommendations.** There are many inhibitors on the market today. Some are single chemicals and others are combinations of several chemicals. Some are liquids and others are solids. The suppliers will readily make recommendations regarding the types of inhibitors and injection methods based on the information supplied by the designer.
2. **Document the results of the investigation.** After a decision has been made as to the need for an inhibitor, the details should be carefully documented. The kind and amount of inhibitor and the frequency and method of addition should be reported. Of course, there are many ways to report these. However, make sure that this information is included in the Overall Materials Guide discussed in Chapter 5.8 with an explanation of why an inhibitor was needed and why the specific inhibitor was selected.
3. **Monitor the results of the inhibitor systems.** (The monitoring system should be specified by the designer *before* operation.) Monitoring can be done using analytical techniques, corrosion monitoring instruments, and process equipment inspection. Management systems to assure long-term maintenance application should also be installed.

☐ 6.6 Protective Coatings

The use of protective coatings is one of the main methods of corrosion control available to the designer. This type of protection is particularly important for carbon steels, as has been mentioned previously. There are many types of protective coatings available, including:

- Organic coatings (paint films)
- Anodizing aluminum
- Conversion coatings
- Electroplating
- Other metallic coatings
- Flame spraying
- Glass coatings
- Clad metals

Organic Coatings (Industrial Coatings — Paint Films)

Of the above listed coatings, probably organic coatings are used most, and for that reason, they will be given the most emphasis in this section. When the designer contemplates using a protective coating, he should remember that the proper selection of an organic coating is only part of the job. Of equal importance is the preparation of the metal surface that is to be protected. The coating itself is usually in three parts, the prime coat, the intermediate coat, and the top coat. This constitutes a coatings system along with the surface preparation. The coating can fail if the surface preparation is not done properly and, conversely, if the surface preparation has been properly done but the coating is wrong for the process environment, failure can also occur.

Surface preparation. Proper surface preparation is probably the most important factor of the coating system. Even the most resistant coating will fail if the surface has not been properly prepared, as mentioned above. There are many ways of doing this, depending on the nature of the surface to be cleaned. A few of these methods are:

- Abrasive cleaning (sand or grit blasting)
- Steam and water cleaning
- Flame cleaning
- Pickling
- Solvent or vapor degreasing
- Wire brush (manual or power)

Wire brushing is probably the most common method of preparing surfaces and is the least effective. It is also probably the most inexpensive method. Abrasive cleaning using either centrifugal or air pressure blasting is considered to be the most effective cleaning method.

The Steel Structure Painting Council (SSPC) has established specifications[63] for surface prepartion of various types of protective coatings. Three of their most used specifications are:

SSPC-SP5 — White Metal Blast Cleaning; used for inorganic zinc coatings.
SSPC-SP6 — Commercial Blast Cleaning; used for alkyd, epoxy ester, coal tar epoxy, and chlorinated rubber coatings.
SSPC-SP10 — Near White Blast Cleaning; used for phenolic, epoxy, vinyl, and organic zinc coatings.

Specific surface preparation specifications have also been published in NACE Standards TM0170-70 and TM0175-75 ("Visual Standard for Surfaces of New Steel Airblast cleaned with Sand Abrasive" and "Visual Standard for Surfaces of New Steel Centrifugally Blast Cleaned with Steel Grit and Shot" respectively).

When blasting, a pattern of hills and valleys (as seen in magnified cross sections) is created on the surface being cleaned. If the peaks are too high, spot-rusting may occur in service. If the valleys are too shallow, coating adhesion may be sacrificed. The *profile* or *anchor pattern* of the blasted surface generally should not exceed one half of the dry film thickness of the first coating to be applied.[64] Surface preparation should be included in the designer's protective coating specification and purchase order.

Sand or grit blasting is favored as the most dependable over the other cleaning methods, as mentioned before. However, in all cases, the advice of the formulators or vendors of the protective coatings should be followed in regard to specific surface preparations required for their coatings.

Prime coat. The purpose of the prime coat is to bond the intermediate and top coats to the metal surface. The prime coat is a very important factor in the service life of the coating systems. The most common primer used in industry is the inorganic zinc primer. This primer acts as a cathodic protector to the base steel and is considered the most efficient of the primers available. There are other primers called *wash primers* and *special primers* such as coal-tar and chlorinated rubber.

Whatever primer is specified, the designer should remember that it must be compatible with the intermediate or top coat. Usually, primers are 3 mils or less in thickness.

Intermediate and top coats. The intermediate or second coat (when used) is usually the same as the top coat. However, in some applications, it can be different from the final top coat. Its main purpose, other than film buildup, is the tying-in of the prime coat to the top coats. The top coat mainly controls the corrosion resistance of the coating system. This is the coating that commands the most interest for the designer.

There are a myriad of organic coatings available for the designer to specify. These coatings include:

Acrylics	Oils, greases, and waxes
Alkyd	Phenolics
Asphalt	Polyesters
Chlorinated rubber	Polyvinyl chloride acetate
Coal tar (epoxy, urethane)	Urethane
Epoxy	Silicone
Epoxy ester	Vinyl
Fluorocarbons	Vinylidene chloride
Furfuryl alcohol	Zinc-filled *inorganic*
Neoprene	

There is a great deal of information available concerning the selection of protective coatings such as the NACE *Coatings and Linings Handbook*,[65] *Coatings and Linings for Immersion Service*,[66] and *Corrosion Prevention by Protective Coatings*.[67]

Table 6.2 is included to help the designer in selecting an appropriate coating.[68] It can be used to determine which coating systems should be tested for the process conditions expected. The generic name of the coating is listed on the left side of the table. The designer should keep in mind that some formulations on the market are combinations of several coating materials (although named for the predominant generic coating) and may not show all the characteristics (good or bad) noted in the table. However, resistance of the specific coating to various environments along with temperature limitations noted in the table should aid in the designers selection.

The NACE reports listed on the far right of Table 6.2 can be obtained from NACE. They give detailed instructions on the application of specific industrial coatings. These reports are also included in the NACE manual and handbook referenced above.

Designing for protective coatings. The coating can also fail if the design of the surface to be coated has not been done properly. To assure that the design of the structure or part is amenable to coating, the following steps should be taken:

1. Eliminate sharp corners.
2. Use butt welding instead of lap welding.
3. Remove weld spatter.

TABLE 6.2 — Characteristics of Various Common Industrial Coatings[1]

Generic Name	Surface Prep.[2]	Thickness, mils	Brush	Spray	Dip	Electrostatic	Fluidized Bed	Roller	Abrasion	Acids	Alkalis	Gases	Oil & Fats	Oxidizers	Solvents	Sunlight	Temp. Limit, °F	Water Fresh	Water Salt	Toxic	NACE Report Number
Alkyd	3	4.5-5	X	X	—	—	—	X	F	F	P	F	G	P	G	E	218	G	G	No	6B165
Asphalt	3	2.5-250	X	X	X	—	—	—	F	G	P	G	P	P	P	E	300	G	—	No	6A166
Chlorinated Rubber	3	5	X	X	X	—	—	X	F	G	G	G	G	G	P	P	300	G	G	No	6A356
Chlorosulfonated Polyethylene	1	8	—	X	—	—	—	—	E	F	E	G	G	E	P	E	140[3]	E	E	No	6A166
Coal Tar (Epoxy, Urethane)	2	30	X	X	—	—	—	X	E	E	E	E	E	E	F	E	212	E	E	Yes	6A162
Epoxies	2-3	7	X	X	X	X	X	X	G	E	G	F	G	E	E	G	250	F	F	No	6A256
Fluorocarbons	1	7	—	X	X	—	—	—	—	G	G	E	G	E	E	E	390	—	—	—	6A163
Furfuryl Alcohol	1-2	20	X	X	X	—	—	X	W	G	G	G	E	G	E	F	200	G	—	No	6A263
Neoprene	1	20	X	X	—	—	—	—	E	G	G	G	E	P	G	E	200	E	E	No	6A363
Phenolics[5]	1	5-7	X	X	—	—	—	—	F	G	P	G	E	P	G	G	300	G	E	No	6A362
Polyesters[4]	1	—	—	X	—	—	—	—	—	G	F	G	E	P	G	G	175	G	E	Yes	6A462
Polyvinylchlor-Acetate	1	250	X	X	X	—	—	X	G	F	G	P	E	F	F	E	150	E	E	No	6A154
Urethane	1	12	X	X	—	—	—	—	E	P	P	G[5]	E	P	P	G	250	E	E	No	6A262
Vinyl	1	8	X	X	X	—	—	X	G	E	E	E	E	E	P	E	150	E	E	No	6B163
Vinylidene Chloride	2	10	X	X	X	—	—	X	E	G	G	F	G	F	F	G	150	E	E	No	6A157
Zinc-Filled Inorganic	1	4	—	X	—	—	—	—	E	P	P	—	E	P	E	E	700	E	E	Yes	6B161

[1] Data are adapted from information available in the NACE Technical Committee reports referenced herein. Not all data match exactly the criteria in the reports. Refer to reports for exact and complete information.
[2] Numbers refer to specific surface preparation grades included in NACE Technical Committee Report No. 6G153.
[3] Wet temperature maximum.
[4] Random reinforced with chopped glass fibers.
[5] Baked formulations

LEGEND: E = Excellent, G = Good, F = Fair, P = Poor, X = Used, — = Not Used. Recommendations are advisory only. Where no recommendations appear, data are lacking.

4. Avoid designs that will collect or hold water and debris.
5. Provide drainage in recessed zones.
6. Specify the removal of rough surfaces by sanding, grinding, or other means.
7. Specify that welds are to be continuous, no skip welding.
8. Specify the rounding of all corners.
9. Eliminate hard to reach places.
10. Specify that rivets be caulked.
11. Specify that surfaces be easily accessible.
12. Provide a continuous and even surface to allow complete bonding of the coating to the metal surface.
13. Eliminate lattice construction or other intricate construction that would be difficult to coat. Instead, use pipe construction or other simple designs.
14. Specify that irregular surfaces, such as exposed threads of pipe, should be cleaned and caulked. Addition thickness of the coating should also be applied.
15. When a thick plate is to be joined to a thinner plate, design the surface to be coated to be flat, as shown in Figure 6.22.
16. Relocate internal stiffeners or structural supports for tanks on the outside of the tank, leaving the inside surface smooth. If the stiffeners are located inside the tank, the coating will be difficult to apply and may fail prematurely.
17. Do not incorporate crevices or pockets in the design. Where crevices cannot be avoided, specify that they be welded and welds be ground prior to the application of the coating.
18. Ascertain that mill scale has been removed from the steel surface. Mill scale is cathodic to steel and may spall, taking the coating with it.
19. When designing tanks, vessels, or piping, provide easy access for coatings and inspections. (See Figure 6.23.)[69]

Testing protective coatings. In specifying protective coatings for critical conditions, it is essential that corrosion tests be conducted. The various producers of the protective coatings should be contacted to get their recommendations as to which of their products would be most suitable for the stated process conditions, and then coated specimens can be made up for testing purposes following the vendors instructions. Also, many protective coatings producers will supply steel specimens already coated with their products to be used for testing. Field test panels, such as KTA (KTA-Tator, Inc., Pittsburgh, Pennsylvania) panels, are designed to evaluate certain conditions, such as corrosion

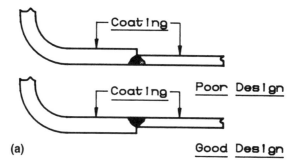

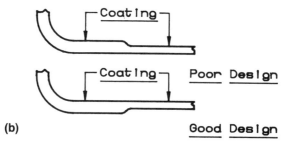

FIGURE 6.22 — (a) Joining unequal thicknesses by welding; (b) unequal thickness in forged head design.

of the base metal, loss of adhesion (particularly at scribe marks), and holes, blistering, and cracking. ASTM Standard D1014 concerns field specimens and gives advice for field testing.

Laboratory apparatus used for testing protective coatings under immersed conditions are shown in Figures 6.24 and 6.25. Each tester consists of a glass tee with two flanges. Two flat steel specimens of the same surface preparations, and prime, intermediate, and top coats that are being considered for service are mounted as blind flanges on the tester with the coating on the inside. The tester is filled with the process solution, while a heating element controls the desired temperature. Figure 6.26 shows a protective coating specimen that has been tested and failed in the apparatus shown in Figures 6.24 and 6.25.

When the coating is to be used for outside service, the specimen should obviously be tested outside. For testing of coatings for equipment to be located in a chemical plant, these specimens should be

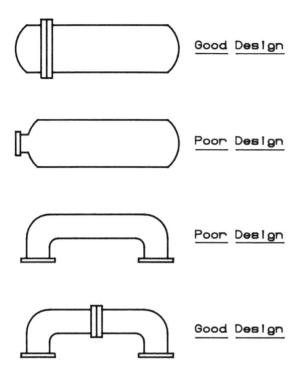

FIGURE 6.23 — *Designs for interior coating.*

located as close as possible to where the equipment will eventually be installed.

Anodized Aluminum

This form of protective coating on aluminum is formed when the aluminum part is made the anode in an acid bath with a lead cathode. These baths are usually made up using either chromic acid, sulfuric acid, or a combination of sulfuric and oxalic acid for so-called "hard" coatings. The reaction at the anode is oxidizing converting aluminum to aluminum oxide (Al_2O_3). Aluminum and aluminum alloys always have a very thin layer of oxide on their surfaces; however, the anodizing process builds up this layer to 0.20 to 0.70 mils in thickness for chromic and sulfuric acid treatments and for the hard coating, the thickness can be as much as 4 mils.

(a)

FIGURE 6.24 — *Protective coatings testers being assembled.*

(b)

FIGURE 6.25 — *Several protective coatings testers being operated; note the reflux condensers.*

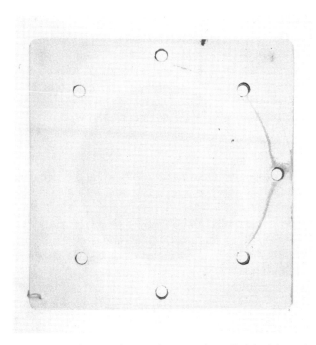

FIGURE 6.26 — *A protective coatings specimen that had been tested (shown at about half size).*

The corrosion resistance of anodized aluminum depends to a great extent on a final sealing operation. The sealant is generally acidified water held at closely regulated temperatures [chromic acid bath 175 ± 2 F (79 ±1 C) and sulfuric acid bath 200 to 212 F (93 to 100 C)] for specific periods of time and pH readings. Sealing changes the anhydrous form of aluminum (Al_2O_3) to aluminum monohydrate ($Al_2O_3H_2O$) called *boehmite*. The coating must be continuous, adherent, and free of surface blemishes for the sealing operation to be effective.

Anodizing, including sealing, significantly improves the corrosion resistance of aluminum, especially in the atmosphere and in seawater. Anodizing also increases the adherence of protective coatings applied to aluminum. It also permits subsequent plating to be effectively applied and increases abrasion resistance. The designer can specify many of the commercial alloys of aluminum to be anodized. However, regular anodizing should not be applied to alloys of more than 5% Cu content and hard coatings to alloys of more than 3% Cu and 7% Si content; otherwise, excessive pitting may result.

When the designer wishes to specify an anodized coating on aluminum or aluminum alloys for a critical service, such as in a marine atmosphere, he should note in the purchase order that the coating must be produced according to the requirements of military specification MIL-A-862S or to those of other pertinent specifications. The designer should also specify that the final coating is to be evaluated for corrosion resistance in accordance with ASTM Specification B-117-85 ["Standard Method of Salt Spray (Fog) Testing"].

Conversion Coatings

Chemical conversion coatings are formed by a chemical oxidation-reduction reaction of the surface of a metal with a suitable chemical solution. Conversion coatings are popular because no electric current is required to produce them. Immersion, spraying, wiping, brushing, or other such wetting methods may be used and are less expensive than electrolytic methods, such as anodized coatings, which are formed, as noted above, by an electrochemical reaction.

Phosphate coatings. Phosphate coatings are applied to most grades of steel and to aluminum alloys. Phosphate coatings on aluminum are usually interchangeable with anodized coatings. Phosphate coatings on steel are formed when dilute phosphorous acid contacts the surface of the steel, which reacts to form an integral layer of crystalline phosphate. This layer is insoluble in water and is moderately protective. The coatings are about 0.1- to 2-mils thick, although the coating is usually referred to by weight of phosphate per square foot.

The principle phosphate coatings are zinc, manganese, and iron. The main application of all three coatings is as a base for protective coatings. Zinc phosphate, because of its zinc content, has an increased rust proofing ability; the iron phosphate coating is used as a bonding base for nonmetals to steel. The zinc phosphate coating is probably of the most interest to the designer. However, it should be remembered that the zinc phosphate coating alone is only resistant to mild corrosion for short spans of time.

Chromate coating. Chromate conversion coatings are formed on metal surfaces when they are contacted through immersion or spray by aqueous solutions of chromic acid. The coatings consist of complex compounds of metal and chromium. Many metals and electroplates may be protected by this type of conversion coating, including magnesium, cadmium, zinc, and aluminum. The chromate coating is also used to seal or supplement phosphate, oxide and other nonmetallic protective coatings.

Chromate coatings are very slowly soluble in water; therefore, this coating will provide variable protection in this media, depending upon the thickness of the coating. When the coating thickness is adequate, it will significantly retard corrosion in rural and sea coast environments. However, these coatings are most useful in providing a good non-porous bond for paints or protective coatings.

Metal oxides. For many years, steel gun parts have been "blued" by exposing the parts to dry steam and alkalies. In later years, chemical baths were developed to produce the same conversion coating. The coating consisted of a very thin film of iron oxide, which is predominantly Fe_3O_4. The corrosion resistance of this coating can be enhanced by impregnation with lacquers, waxes, or oils. Metal oxide coatings may also be applied to other metals, such as copper and nickel. Such coatings are applied by hot alkalies or controlled thermal oxidation.

Sealing anodized or conversion coatings. When the designer specifies anodizing or a conversion coating, he should make sure that an additional treatment is added to assure the integrity of the surface. Anodized surfaces, for instance, should be sealed. Phosphatizing should be either sealed or have a protective coating applied over it. Chromatizing is seldom used alone, but is used as a coating base. The bluing on steel must be carefully oiled and maintained to adequately resist corrosion.

An example of a conversion corrosion coating on aluminum that did not hold up without an additional sealing treatment is shown in Figures 6.27 and 6.28. Figure 6.27 shows four photomicrographs of the same area on an aluminum specimen at 200, 500, 1000, and 2000X using a scanning electron microscope (SEM), thereby disclosing a very porous coating cross section. When this coating was exposed to an industrial atmosphere containing ferric chloride, it failed in a short time. Figure 6.28 shows the cross section of the same area of a part exposed to this industrial atmosphere at 200, 500, 1000, and 2000X. It is obvious that the substrate was severely attacked by the atmosphere because the coating had not been properly applied and sealed.

Electroplating

Electroplating is accomplished by making the part to be plated the cathode, while metal ions from an anode at high current densities are transported to the cathode where they are reduced and plated on the cathode ($M^{+n} + ne \rightarrow M^0$), as covered in Chapter 1.4. The electrolyte or plating bath (for general purpose plating) can be of several acid compositions, such as boric acid combined with nickel sulfamates and

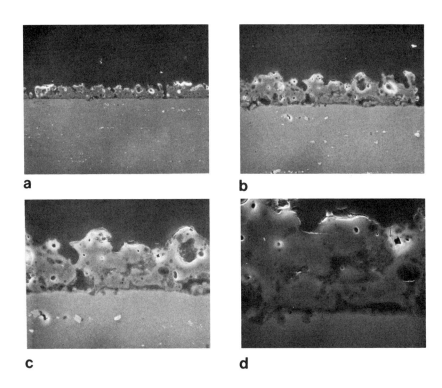

FIGURE 6.27 — *Very porous conversion coating over aluminum; scanning electron micrographs at (a) 200, (b) 500, (c) 1000, and (d) 2000X magnifications.*

in other baths with nickel fluoroborate.

Since the solutions above are acidic, hydrogen can be formed at the anode. Nickel is somewhat permeable to hydrogen; therefore, embrittlement of the steel substrate can occur if, after electroplating, the plated part is not baked at about 375 F (about 190 C) for one to three hours. (Refer to Chapter 3.9.)

Nickel plating. Nickel plating was the first plating used by industry. This type of plating is used primarily to protect alloys of copper, zinc, and iron against marine, rural, and industrial atmospheres. There are great variations in baths and procedures for the plating of nickel. The coating thickness can vary from 0.2 to 2 mils.

Chromium plating. *Nickel* plating takes on a yellow tarnish during long exposure to mild corrosive conditions and a green color when corrosive conditions are more severe. Chromium plating, which was

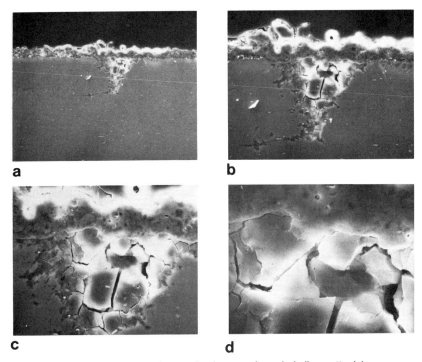

FIGURE 6.28 — *Conversion coating improperly sealed allows attack in an industrial atmosphere; scanning electron micrographs were taken at (a) 200, (b) 500, (c) 1000, and (d) 2000X magnifications.*

introduced approximately seventy years ago, was found to be an ideal plating over nickel plates, stopping such corrosion.

For exterior use, such as automobile bumpers and boat hardware, a protection system of copper plate 0.2 mils; nickel plate around 1 mil and chromium plate 0.01- to 0.03-mils thick have proven successful.

The so-called "hard chromium" plating, known as industrial or engineering chromium plate, is usually thicker than normal chromium plate. The thickness can be 0.10 to 20 mils. This coating is generally applied directly to the base metal and ground to a finished dimension. Hard chromium plate is also used to salvage under-size parts. Hard chromium-plated rolls are extensively used in the chemical industry for various production functions.

Cadmium plating. Cadmium plating is frequently used to protect steel against corrosion. Since cadmium is anodic to iron, it acts as a cathodic protector. This plate is usually applied to a thickness of about 1 mil and is intended to withstand atmospheric corrosion. Because of its ability to minimize galvanic corrosion, it is frequently used between dissimilar metals. The plating baths used are usually composed of cyanide or fluoroborate solutions. Cadmium plating is inexpensive, especially for small parts that can be plated in a barrel or automatic plating equipment. Springs are often electroplated with cadmium for protection against corrosion and abrasion.

Zinc plating. Zinc plating is also anodic to iron and, like galvanized coatings, acts as a cathodic protector even though there may be breaks in the coating. Zinc is applied in thin films of 0.3 to 0.5 mils and is also inexpensive. The most common plating bath is a sodium cyanide solution.

The application of zinc plating is many and varied. The life of a zinc plate in the atmosphere varies directly with the thickness of the plate. Zinc-plated steel is not normally used for process equipment that is continually immersed in aqueous solutions. It is also not used in contact with food or beverages.

Zinc plating requires about the same production equipment as cadmium plating. In mild corrosive conditions, such as rural atmospheres, both zinc and cadmium plating of equal plate thickness have about the same corrosion resistance. However, zinc is superior to cadmium in industrial atmospheres, while cadmium is superior in marine atmospheres. Zinc plating forms a larger amount of corrosion product than cadmium. This is especially important at or near the sea where close-fitting parts like hinges must operate. Therefore, cadmium plating is preferred at such locations.

Other Metallic Coatings

Other metallic coatings include hot dip galvanizing, welding, and flame spraying, as described below.

Hot dip galvanizing. *Hot dip galvanizing* is a method of producing a thick coating of zinc by immersing steel in a molten bath of zinc maintained at temperatures ranging from 830 to 870 F (443 to 465 C). The thickness of the coating varies greatly with bath temperature, immersion time, and withdrawal rate from the bath. It can range up to 4 to 7 oz. per square foot (1220 to 2135 grams/m^2) for steel sheet counting both sides. The coating actually is a coating of several distinct layers. The layers closest to the steel surface are composed of iron-zinc compounds, while the outer layers consist mainly of zinc. The best steels for galvanizing are those containing less than 0.15% carbon. Cast iron can be galvanized but should be low in silicon and phosphorus to avoid brittleness in the zinc-iron layer closest to the surface of the cast iron.

There are many uses for galvanized steel, including many mill products, fasteners of all descriptions, pipes and fittings, structural members, heat exchanger coils, highway guard rails, etc. The zinc on the surface of steel, like zinc and cadmium platings, cathodically protects the underlying steel, even if breaks occur in the coating. The corrosion protection is best in atmospheres that do not contain sulfur gases and other industrial pollution; however, galvanized steel is widely used in industrial atmospheres because galvanizing is so economical. Galvanized steel functions well in marine atmospheres. Although thousands of tons of uncoated galvanized steel are used in this country, its corrosion resistance can be materially enhanced by the application of conversion coatings, such as phosphate, followed by an organic coating. A zinc primer may also be used instead of the conversion coating; however, the phosphate treatment is considered the best. The life of galvanized steel is predicted directly from the thickness of the coating. The designer should specify appropriate ASTM specifications to assure that an effective application job and thickness of coating will be delivered on the product ordered.

Welding. Overlays of protective weld metal over carbon steel have saved many thousands of dollars. Overlays inside steel tanks, on rolls of various kinds and on flanges, are examples of what can be done. Manual welding and automatic welding are often used. Overlays of stainless steels, nickel-base alloys, and almost any weldable metal have also been used for this purpose.

A good example of protective overlay welding was when a circular flange some 40 ft (13 m) in diameter, 1.5-ft (0.46-m) wide and 2-in. (51-mm) thick was specified to be made of AISI 304L stainless steel. Solid stainless steel would have been extremely expensive. Instead, a forged flange made of SAE 516 Grade 70 steel was procured. An overlay of a 1/2-in. (12.7-mm) thickness was made using an AISI 309L stainless steel weld deposit. The electrode of the automatic welder was in the form of a ribbon about 2.5-in. (63.5-mm) wide and about 1/16-in. (1.6-mm) thick to assure uniform overlay. A sample of the weld deposit showed a carbon content on the surface to be under 0.03% carbon. After the welding was completed, the overlay was ground to final dimension.

Flame spraying. This coating method creates a deposit on a substrate by heating either wire or powder to a high heat and then propelling the molten or semimolten drops to the work. Upon impact, the drops flatten out and then adhere to the surface, creating a lamellar structure. When powder is used, a gas is required to propel the heated powder to the equipment to be coated. Oxyacetylene may be used as the heating source. Also used are spray guns using electrically generated plasma to create a very high heat, high enough to melt all known metals. In some applications, such as for high heat and abrasion, the coatings are fused after the spraying has been completed.

The surface of parts must be cleaned thoroughly and roughened to assure a good spray coat. The coatings can be built up to almost any thickness. Thicknesses of up to 1/8 in. (3.2 mm) and more are not unusual. The thickness of built-up shafts, depending on diameter, can vary from 0.015 to 0.040 in. (0.381 to 1.02 mm).

Uses include repair buildup of worn parts, mill roll journals, high-temperature-resistant coatings for rockets and jet aircraft, all types of shafts, including pump shafts (where a stainless spray coat is applied for protection over a high-strength steel) automotive exhaust mufflers, and packing sleeves for centrifugal pumps.

Clad Metals

Cladding of steel with various other metals and alloys has been used extensively in the chemical industry. These metals include stainless steel of all types, nickel, nickel alloys, copper, and tantalum. The steel provides the strength, and the cladding assures specific corrosion resistance. Applications are many and varied, such as storage tanks, pressure vessels, melting pots, and reactors. Clad metals are made by hot rolling the steel and the cladding together or by explosive cladding. The latter type of cladding is performed by a unique process using explosives to drive the pieces together. In each case, a metallurgical bond is formed between the cladding and the steel backing plate. Other metals can also be used as the backing plate to satisfy special requirements.

Welding of clad plates together can be done without any loss of corrosion resistance as follows. The weld groove is made according to the welding design required, i.e., a V- or U-groove. After tacking the two plates together, the carbon steel backing side is welded using carbon steel electrodes or filler metal. The joined plates are then turned over, and the stainless steel side at the joint is gouged out well into the carbon steel side. This groove is then welded with stainless steel electrodes completely filling the gouged-out groove and the stainless steel side.

Classification of Metallic Coatings

Metallic coatings can be divided into two groups. The *anodic group* encompasses coatings that are more active than the metal being coated such as zinc, aluminum, and cadmium on steel, and the *cathodic group*, which includes those where the coating is less active than the metal being coated such as nickel, copper, and chromium on steel. As covered in Chapter 3.2, the anodic group will protect the underlying metal even though it has scratches and holidays. Conversely, cathodic coatings may *accelerate* corrosion at breaks or pinholes in the coating. The designer should therefore specify that extra care be exercised when handling cathodic coatings to avoid marring the surfaces.

Glass Linings

Glass fused onto metals forming coatings are widely used today in the chemical and food industries, as well as others, because of resistance to a wide variety of chemicals and corrosive solutions. One of the major uses for glass linings is in the handling of strong acids (hydrofluoric acid is the exception). Glass linings are not resistant to hot concentrated bases. A definite disadvantage of glass linings is, of course, brittleness and easy breakage. However, small breaks and pinholes can be repaired by tantalum plugs. Products that will adhere or stick to walls of metal equipment will not adhere to glass linings in many instances. This is a definite advantage of using glass linings. Glass-lined piping reactors, storage tanks, and many other pieces of equipment are available to industry.

Glass linings are produced as follows: (1) a special grade of steel is selected that has a low gas content, thus reducing the amount of gas that can be expelled during firing. (Escaping gas can cause pinholes to be formed in the glass lining.) (2) The steel is sandblasted and/or pickled for cleanliness. (3) Powdered glass is spread evenly over the steel surface and heated to 1400 to 1600 F (760 to 871 C). (4) The glass flows and bonds to the steel. Generally, two types of glass are used. A borosilicate glass provides the best bond to the steel. This is followed by a high-silica glass that has better acid resistance.

The designer should contact glass lining companies for advice on types of glass thicknesses, etc., for specific applications.

Actions the Designer Should Take when Considering Using Protective Coatings

The following list describes the actions a designer should take when considering protective coatings:

1. Ask whether it is less expensive, considering maintenance cost as well as installation cost, to use a steel protected by a coating rather than a more expensive metal or alloy that is resistant to the process conditions. See Chapter 1.2 for determining annual costs.

2. If a protective coating is decided upon, make sure that it is adequately tested.

3. Make your specification clear as to what is required. An example of a specification for an *organic coating* might be:
   - Steel surfaces shall be grit blasted to white metal following SSPC-SP5.
   - Prime coat shall be a commercial inorganic zinc (NACE Report 6B161) 1-mil thick, min.
   - Top coats shall be Smith Co. Hi-Build Epoxy (amine cured) No. 72.
   - Coverage shall be applied in two coats for a final dry film thickness of 7 to 8 mils.
   - All process and safety precautions of the producers shall be carefully followed.
   - Inspection shall be conducted after surface preparation; prime coat and finish coats have been completed. Thickness shall be verified, and the surface of the coating carefully examined for holidays and pinholes.

4. When specifying an electroplate, the designer should remember that the quality of an electroplate can depend on the design of the part to be plated. For instance, have all sharp corners of the part removed before plating and round off all edges. It is best to avoid concave surfaces and (when possible) flat surfaces. Design surfaces to be slightly crowned or convex. The reason for this is that the latter curvature fosters a more even plate with less buildup on the edges of the part.

■ References

1. R. Tunley, "Time Bomb in our Tap Water," *Readers Digest,* Jan. 1985.
2. F. A. Rohrman, "Corrosion Studies for the Petroleum Industry," *Petroleum Refiner,* June 1947.
3. C. P. Dillon, "Economic Evaluation of Corrosion Control Measures," *Materials Protection,* Vol. 4, No. 5, p. 38, 1965.
4. NACE Standard RP0272-72, "Direct Calculation of Economic Appraisals of Corrosion Control Measures," National Association of Corrosion Engineers, Houston, TX, 1972.
5. G. Sorell, "Controlling Corrosion by Process Design," *Chemical Engineering,* July 19, 1968, p. 162.
6. R. J. Landrum, "Designing for Corrosion Resistance, Part I," *Chemical Engineering,* Feb. 24, 1969, p. 120.
7. H. H. Uhlig, *Corrosion Handbook,* J. Wiley & Sons, Inc., New York, NY, 1948.
8. Mackay and Worthington, *Corrosion Resistance of Metals and Alloys,* Reinhold, New York, NY, 1936.
9. M. Henthorne, "Polarization Data Yield Corrosion Rates," *Chemical Engineering,* July 26, 1971, p. 99.
10. J. A. Collins, M. L. Monack, "Stress Corrosion Cracking in the Chemical Process Industry," *Materials Protection and Performance,* Vol. 12, No. 6, p. 11, 1973.
11. R. J. Landrum, "Evaluation of Structural Materials for Corrosion Resistance," *ASM Engineering Quarterly,* Vol. 1, No. 2, p. 45 (May 1961).
12. R. J. Landrum, J. Teti, "Apparatus for Corrosion Testing," US Patent 3228, 236, E. I. du Pont de Nemours & Co., Inc., dedicated to the public, 1969.
13. N. D. Groves, "Hot Wall Tester," *Metal Progress,* May 1959.
14. ASTM offers a directory of testing laboratories, featuring 1000 laboratories listed by detailed subject and alphabetical indexes. For more information, contact ASTM Customer Service, 1916 Race St., Philadelphia, PA 19103.
15. L. D. Yates, E. D. Tait, "Nomograph for Calculation of Corrosion Rates," *Metal Progress,* Dec. 1942.

16. T. F. Degnan, private communication, March 1982.
17. J. L. McPherson, "How Good Design Controls Corrosion," *Plant Engineering,* Jan. 1957, p. 102.
18. W. B. DeLong, "Testing Multiple Specimens in a Modified Boiling Nitric Test Apparatus," presented at the 52nd Annual Meeting of ASTM, ASTM, Philadelphia, PA, June 1949.
19. "British Navy Board Report to the Admiralty of the First Coppering Experiment, August 31, 1763," reprinted in *American Neptune,* July 1941.
20. "Design and Installation of 90-10 Copper-Nickel Seawater Piping Systems," International Nickel Company, Inc., New York, NY.
21. K. G. Compton, A. Mendizza, W. W. Bradley, "Atmospheric Galvanic Couple Corrosion," *Corrosion,* Vol. 11, No. 9, 1955, p. 383t.
22. "Aspects of Galvanic Corrosion," *International Nickel Co. Corrosion Engineering Service,* International Nickel Co., Inc., New York, NY, 1950.
23. R. J. Landrum, "Designing for Corrosion Resistance, Part II," *Chemical Engineering,* March 24, 1969, p. 172.
24. B. Cohen, "Designing to Prevent Aerospace Weapon Corrosion," *Materials Performance,* Vol. 15, No. 9, p. 23, 1976.
25. M. G. Fontana, "Corrosion," *Industrial and Engineering Chemistry,* Vol. 40, p. 95A (May 1948).
26. Courtesy of F. M. Allen and the Savannah River Plant, 1984.
27. T. Smith, "Reducing Corrosion in a Heater Plant," *Anti-Corrosion,* 1983.
28. R. E. Tatnall, Annual Conference on Great Lakes Research, State University of New York, May 27, 1983.
29. R. E. Tatnall, "Fundamentals of Bacteria Induced Corrosion," *Materials Performance,* Vol. 20, No. 9, p. 32, 1981.
30. J. R. Postgate, *The Sulfate Reducing Bacteria,* Cambridge University Press, Cambridge, England, 1979.
31. G. Kobrin, "Corrosion of Microbiological Organisms in Natural Waters," *Materials Performance,* Vol. 15, No. 7, p. 38, 1976.
32. B. F. Brown, "Stress Corrosion Cracking Control Measures," NBS Monograph No. 156, 1977.
33. R. W. Staehle, *Stress Corrosion Cracking and Hydrogen Embrittlement of Iron Base Alloys,* NACE, Houston, TX, 1977.
34. J. Q. Lackey, private communication, 1983.
35. From literature of the Stellite Division of Haynes International, Inc., Kokomo, IN.

36. G. E. Moller, "Designing with Stainless Steels for Service in Stress Corrosion Environments," *Materials Performance*, Vol. 16, No. 5, p. 32, 1977.
37. A. W. Dana, W. B. DeLong, "Stress-Corrosion Cracking Test," *Corrosion*, Vol. 12, No. 7, p. 309t, 1956.
38. *Corrosion Basics: An Introduction*, A. deS. Brasunas, Ed., p. 352, 1984.
39. K. F. Krysiak, "Unusual Metal Failures in Chemical Plant Equipment," *Materials Performance*, Vol. 22, No. 11, p. 23, 1983.
40. From several tests conducted by J. M. Stone and D. Warren.
41. A. V. Allessandria, "Refineries Report New Cases of Stainless Steel Failures," *Petroleum Refiner*, Vol. 40, p. 151 (May 1960).
42. S. L. Pohlman, T. N. Andersen, "Acid Car Corrosion Mechanism, Monitoring and Protection," CORROSION/87, Paper No. 24, NACE, Houston, TX, 1987.
43. S. K. Brubaker, "Materials of Construction for Sulfuric Acid," 25th Annual Liberty Bell Corrosion Course, NACE, Houston, TX, Oct. 1987.
44. M. Tiivel, F. McGlynn, "Avoiding Problems in Sulfuric Acid Storage," presented at the American Institute of Chemical Engineers Meeting, American Institute of Chemical Engineers, New York, NY, 1986.
45. Courtesy of the American Welding Society, Miami, FL.
46. H. A. Norman, private communication, July 28, 1975.
47. Data from "Caustic Soda Service Chart," *Corrosion Data Survey*, NACE, Houston, TX, 1960.
48. E. C. Huge, E. C. Piotter, "The Use of Additives for Protection of Low Temperature Corrosion in Oil-Fired Steam Generating Units," ASME Report No. 77.4, American Society of Mechanical Engineers, New York, NY, April 1955.
49. S. R. Coburn, "Minimizing Corrosion through Proper Design," Liberty Bell Corrosion Course, 1967.
50. *Seawater Corrosion Handbook*, M. Schumacher, Ed., Noyes Data Corp., Park Ridge, NJ, p. 36 (Table 34), p. 38 (Table 35), 1979.
51. H. E. Atkinson, private communication, Sept. 1983.
52. A. H. Tuthill, C. M. Schillmoller, "Guidelines for Selection of Marine Materials," Ocean Engineering Conference, Marine Technology, International Nickel Co., Inc., New York, NY, June 1965.
53. Courtesy of the Society of Automotive Engineers, Warrendale, PA (copyright 1982).
54. T. F. Degnan, "Specifications and the Corrosion Engineer," NACE

Regional Conference (Pittsburgh, PA), NACE, Houston, TX, Oct. 6, 1965.
55. *The NACE Book of Standards,* NACE, Houston, TX, 1988.
56. W. C. Rion, R. J. Landrum, *Stainless Steel Information Manual,* E. I. du Pont de Nemours & Co, Inc., Wilmington, DE, April 1968.
57. J. R. Grieve, A. M. Bounds, "Nondestructive Testing of Small Tubing," *Metal Progress,* Dec. 1960.
58. J. C. Bovankovich, private communication, March 1983.
59. *Corrosion Basics: An Introduction,* A. deS. Brasunas Ed., NACE, Houston, TX, p. 179, 1984.
60. O. L. Riggs, C. E Locke, N. E. Hamner, *Anodic Protection: Theory and Practice in the Prevention of Corrosion,* Plenum Press, New York, NY, 1981.
61. F. Flade, "The Passivity of Electrodes," *Z. Physik Chemie,* Vol. 6, 1911.
62. *Corrosion Basics: An Introduction,* A. deS. Brasunas, Ed., NACE, Houston, TX, p. 141, 1984.
63. Steel Structures Painting Council, Surface Preparation Specifications, (ANSI A159.1-1972), Pittsburgh, PA, 1972.
64. J. Lichtenstein, L. F. Flaherty "Basic Corrosion Control Design— Coatings," *Materials Performance,* Vol. 17, No. 7, p. 19, 1978.
65. *NACE Coatings and Linings Handbook* (looseleaf), NACE, Houston, TX, 1985.
66. *Coatings and Linings for Immersion Service,* NACE, Houston, TX, 1972.
67. C. G. Munger, *Corrosion Prevention by Protective Coatings,* NACE, Houston, TX, 1984.
68. *Corrosion Basics: An Introduction,* A. deS. Brasunas, Ed., NACE, Houston, TX, p. 258, 1984.
69. L. T. Hutton, "Design and Fabrication Techniques Used with High Performance Coatings."

■ Appendix
Association Addresses

Aluminum Association (AA)
900 19th St. N.W.
Suite 300
Washington, DC 20006

American Cast Metals Association
(ACMA)
455 State St.
Des Plaines, IL 60016

American Iron and Steel Institute
(AISI)
1133 15th St. N.W.
Washington, DC 20005-2701

American National Standards
Institute (ANSI)
1430 Broadway
New York, NY 10018

American Petroleum Institute (API)
1220 L St. N.W.
Washington, DC 20005

American Society of Mechanical
Engineers (ASME)
345 E. 47th St.
New York, NY 10017-2392

American Welding Society (AWS)
550 N.W. LeJeune Road
Box 351040
Miami, FL 33135

American Water Works Association
(AWWA)
6666 W. Quincy Ave.
Denver, CO 80235

ASTM
1916 Race St.
Philadelphia, PA 19103

Copper Development Association (CDA)
2 Greenwich Office Park
Box 1840
Greenwich, CT 06836-1840

Materials Technology Institute of the
Process Industries (MTI)
12747 Olive Street Road
St. Louis, MO 63141

National Association of Corrosion
Engineers (NACE)
1440 South Creek Drive
P.O. Box 218340
Houston, TX 77218

Society of Automotive Engineers (SAE)
400 Commonwealth Drive
Warrendale, PA 15096-0001

Tubular Exchanger Manufacturers
Association (TEMA)
25 N. Broadway
Tarrytown, NY 10591

Subject Index

A

ABRASION CAUSES
fretting, 101, 102
hardenable stainless vs, 103
lubrication to reduce, 103

ACIDS, AQUEOUS,
concentrations vs Al, 248
copper vs non-oxidizing, 257

ACIDS, ORGANIC
aluminum vs, 247
stainless steels vs, 239
titanium vs, 259

Aeration, effects of, 38

ALKALIS AND SALTS
copper vs, 256
Ni alloy cracking in 315 C, 124
nickel alloys vs, 239
steels, carbon vs, 119, 249

ALLOYING EFFECTS
copper, 256
Cu vs NH_3, 120
stainless steels vs IGC, 150

ALUMINUM
acids vs, 247
anodizing, 321
mercury attack on, 79
properties, types, analyses, 246
tests for heat treatable, 271

ALUMINUM COATINGS
steel grounding wire with, 119
steel wire in atmosphere, 119

Aluminum hydroxide, Cu tubing cracking in, 119

American Petroleum Inst. Standards, descriptions, 224

AMERICAN SOC. MECH. ENG.,
pressure vessel code, 223
welder qualifications tests, 283

American Water Works Assoc., descriptions of standards, 224

AMMONIA
anodic protection vs, 306

copper vs, 117, 120, 121, 256

ACTIVE AND PASSIVE METALS
active-passive reactions, 22
metal and alloy classification, 22

ANNEALING
characterization, 189
cold work, effects on, 189

Anodes, galvanic, types identified, 297
Anodes, inert, identification of some, 296

ANODIC PROTECTION
C steel vs H_2SO_4, 253
characterization, 303
consultants advice on, 309
liquid metals attack, for, 139
reactions during, 23
system design, 307, 309

ANODIZED CONVERSION COATINGS
aluminum, characteristics, 321
military specifications, 324
sealing, alloying limits, 323

Aqueous environments, structures, cathodic protection, 297

ASTM STANDARDS
boiling HNO_3 test for IGC, 154
corrosion acceptance, 222
Huey test, 270
liquid penetrant test, 271
magnetic particle tests, 272
ultrasonic testing, 282

ATMOSPHERIC CAUSES
Al coated steel wire vs, 119
carbon steel vs, 253
cooling tower spray, 204
designing to control, 204
evaluation of, 202
galvanic series in, 68, 72
weathering steel, low alloy, 253

AUTOMOBILES
drainage design, 216
electrical connections, 216
enclosed zones, avoid, 216
geometrical controls, 212
problems related to, 211

soldering practices, 216

B

BACTERICIDES
bacterial attack vs, 113
pressure tests, for water, 114
Barrels, metal, see containers, 78
BIOLOGICAL EFFECTS
biocides to control, 113
copper vs marine fouling, 256
design controls vs, 113
fiber reinforced polymers, 113
materials selection vs, 113
seawater, variations in, 208

Boilers, superheater cracking welds, 128
Bolts, galvanic attack on, 74

C

Cable, nonelectrical, seawater fatigue failure, 142
Cadmium coatings, characterization, 327, 328
CALCULATIONS
corrosion allowance, 44
maximum rate recommendations, 47
plant life factor, 45
rate derivation, determination, 42, 44

Capacitors, Cl concentration effects, 123
Carbides intergranular attack due to, 145

CARBON
graphite attack due to, 78
stainless steel pipes vs, 209

CARBON STEEL
acids vs, 253
alkalis vs, 249
ANSI/ASTM designations, 248
atmospheric exposures vs, 253
contaminated H_2SO_4 vs, 92
end grain attack, 82
galvanizing, hot dip, 328
gases vs, 254
high temperatures vs, 254
industrial use of, 248
inorganic salts vs, 254
seawater velocity vs, 88
steam vs, 254
stress corrosion cracking, 115
stress relief vs NaOH, 191

sulfamic acid cleaning of, 91
sulfuric acid vs, 249
water vs, 253
water, acid vs, 18
welded overlays, 329

CASE HISTORIES OR FAILURE ANALYSES
acid velocity vs pumps, 88
Al alloy tank vs HNO, 13
Al coated steel wire, 119
Alloy 800 vs chlorides, 125
aluminum anodizing failure, 325
auto bearing fretting, 102
bronze grounding wire cracking, 119
buried tank leaks, 210
cannon submerged in ocean, 208
closed heating system, 206
coated stainless steel tanks, 77
coil locations, 196
condenser, AISI 316 tube failures, 131
contaminated H_2SO_4, velocity, 91
copper vs mercury, 121
corrosion fatigue failure, 143
cost factor error, 230
Cu cracking in ammonia, 121
Cu oxide passivation of SS, 311
Damascus steel history, 187
database selection result, 37
end grain attack, 83
failure due to poor drainage, 199
failure statistics, 24
fastener attack by copper, 67
galvanic attack from C, Cu, 79
H grooving, C steel, 159
HCl, sulfamic acid compared, 91
heat exchangers, 11, 12, 97
heat treatment, 192, 193
intergranular attack, 18Cr-8Ni, 146
lead melting pots, 195
liquid metal attack by Zn, 179
Mg containers for HF acid, 78
mixer crevice attack, 63
Monel vs hot amines, 125
paper mill roller shells, 13
pipe materials vs H_2SO_4, 90
pipe replacement, 10
pipe size change economics, 92
pitting, causes of, 104
pitting from sedimentation, 106
pitting from wet insulation, 106
plant location testing, 204
poor drainage vs Al tubing, 198
pump vs corrosion fatigue, 142
PVC cracking of stainless, 135

■ 340 ■

roof, galvanic cell avoid, 73
Savannah river plant tanks, 283
SCC due to welding, 127
specifications, interpretation, 229
specifications, poor, 217
stagnant seawater vs, 95
stagnant water in turbine, 96
stainless drums vs acids, 52
stray current vs SS pipe, 210
sulfamic acid vs C steel, 91
sulfuric acid vs welds, 252
tank bottom failure, 59
tank seam welding, 184
tube cracking by chlorides, 132
unusual instances causing, 77
weld materials compatibility, 177
welded overlays, 329

CAST IRON
contaminated H_2SO_4 vs, 92
graphitization, dealloying, 156
Ni content vs dealloying, 157

CATHODIC PROTECTION
application, uses, limits, 297
characterization, 293
corrosion fatigue vs, 142
immersed structures, 297
liquid metal, not used vs, 139
materials amenable to, 298
pitting controls by, 107
potential survey methods, 301
site survey preceding, 299
specification alternatives, 298
system surveillance, 300
voltage standard, 301

CAVITATION
brass pump impeller vs, 95
characterization, 94

Cells, corrosion, factors in formation of, 17
Chemical industries, economics, 1
Chemical surface preparation, heating following, 140

CHLORIDES
austenitic SS cracking in, 124
concentration controls, 123, 130
gaskets, seals, etc., 134
insulators, adhesives in, 135
localized concentration, 131
nonmetals, vs stainless steels, 135
pitting of stainless steel, 105
temperature effects, 129

thermal insulation concentrations, 132

Chlorine, titanium vs, 260
Chromate conversion coatings, characterization, 324
Chromium coatings, applications, 327

CLEANING,
design to facilitate, 97
pitting control by, 107
precautions during, 134
sulfamic acid vs C steel, 91
titanium, vs high temperature, 137
water systems, 314

COAL
corrosion cell diagram, 14
excellent weld design, 188

COATINGS
books, reports about, 317
characterization, 314
corrosion fatigue vs, 144
decisions on, 332
pitting, control by, 107
preventive for SCC, 137
scheduling prerequisites, 204
surveillance requirements, 205
types available, 317
used vs chlorides on SS, 136
water absorption by, 201

Coatings, metal, anodic vs pitting, 107

COLD FABRICATION
exfoliation due to, 144
grain effects due to, 144

Cold work, effects, annealing controls for, 189
Combustion gases, effects in atmosphere, 202

COMPRESSIVE PRESSURE
corrosion fatigue vs, 141
shot peening vs SCC, 135

Computer software, materials qualification, 269

CONCENTRATION CELLS
crevice attack, 49
decisions modified by, 46
metal ion, 49
oxygen, 49
surface effects causing, 24

■ 341 ■

CONCENTRATIONS OF CHEMICALS
caustic cracking C steel, 118
chlorides, control of, 130
Condensate, aqueous, steam for pressure tests, 114
Consultants, engineers, inhibitor decisions, use of, 311
Containers, stainless steel drums, 52, 57
Contaminantion, atmospheric, effects of, 202

CONVERSION COATINGS
chromate, description, 324
galvanizing, advantages of, 329
oxide, 325
phosphates, suitable alloys, 324

Cooling systems, spray effects, 204

COPPER
alloying effects, 120, 256
ammonia vs alloys, 117
applications, components, 255
casting, weldability, 254
hydrocarbons vs, 256
mercury vs, 117, 121
nitric acid vs, 257
nitrogen compounds vs, 117
oxygen vs, 257
properties, 256
SCC of alloys, 117, 123
sea marine environments vs, 257
velocity effects on alloys, 87

COPPER, WROUGHT
cracking in ammonia, 119
cracking in atmosphere, 119
see copper also, 255

CORROSION ALLOWANCE
calculation of, 44
pitting vs, 107

CORROSION EROSION
characteristics of, 86
design controls for, 97
steam velocity causing, 99

CORROSION PRODUCTS
effects of, 20
influence on passivity, 70
pitting, formation due to, 103
titanium protected by, 257

COUPONS, FIELD TEST
racks for exposure, 287

surveillance, use of, 286

Coupons, laboratory test, heat treatment, 286

CRACKING
austenitic SS superheater, 124
carbon steel, caustic, 118
chlorides vs stainless steel, 123
dye penetrant test for, 271
hydrogen induced, 137
liquid metal, 138
magnetic particle tests, 271
materials vs H-induced, 140
tests for, 271

CREVICES
automobiles, attack control, 212
avoidance in welds, 170
Cr oxide attack on SS, 64
design preventions for, 66
fasteners, attack at, 64
gaskets, other nonmetals, 65
heat exchangers, tube joints, 61
mechanism, 49
mixer attacks, 63
seawater, severity, 210
skip welds, due to, 60
tank bottoms, 56
tests for, 271

Cyclic stresses, corrosion fatigue, influence of, 141

D

Deaeration, fretting controlled by, 103
Deaerators, mechanism, 311

DEALLOYING
alloying vs, 155
brass susceptibility, 155
chromium vs hydrogen chloride, 156
Co by H_2SO_4, Cu-Al alloys, 156
copper alloys susceptible to, 155
Cu-Ni, Ag-Au alloys, 157
Cu-Zn alloys, 157
environments causing, 155
iron from cast iron, 156
materials susceptible to, 155
solder, Si-Cu, Cu-Ag alloys, 157

Decarburizing, corrosion fatigue, avoid, 144

DESIGN
cathodic protection determinants, 299

cleaning, to facilitate, 97
coating, surfaces for, 317
corrosion fatigue vs, 143
end grain attack, to avoid, 83, 85
equipment life factors, 11
fretting reduction by, 102
galvanic attack prevention, 80
general velocity controls, 98
geometrical rules, 205
heat exchange tube orientation, 132
heat exchangers, water box, 99
hydrogen grooving controls, 162
impingement controls, 93
microbe attack, controls, 113
persons involved with, 2
pitting reduction strategies, 107
polymer piping supports, 209
stagnation reduced by, 95
sulfuric acid controls, 251
tight fit avoidance, 129
turbulence reduction by, 98
velocity reduction by, 92, 97
welds, materials compatibility, 177

Dissolved solids, seawater variations, 208

DRAINAGE, MECHANICAL METHODS
automobiles, design for, 216
design for proper, 198
liquids, facilitation of, 197
pipe by correct slope, 201
tank design for good, 199

Drinking water, problems caused by, 205

DUCTS OR PIPES
curve angles controls, 209
turbulence effects, 94

Dye penetrant nondestructive testing, crack, crevice detection, 271

E

ECONOMICS
anodic protection, 309
chemical industry, 1
clogged pipes, 1
coatings, 332
cost calculations, 5
depreciation factors, 7
equipment durability factors, 11
specifications, cost factors, 229
tanks, steel, 1
tax factors, 7

Eddy current testing tube discontinuities, 282
Electrical equipment, automobile, protection of, 216

ELECTRICAL INSULATION
bolts vs galvanic attack, 75
galvanic cells, use, vs, 73

Electrochemical causes corrosion cell, 15

ELECTRODES, WELDING
identification, 283
selection of filler metal, 175

Electronic effects, corrosion cells, 16

ELECTROPLATING, METAL COATINGS
characterization, 325
chromium, 326
heating following, 140
materials, design for, 332
nickel, 326

EMBRITTLEMENT
caustic, carbon steels, 117
hydrogen, 137

END GRAIN ATTACK
carbon steels, 82
designing to avoid, 85
stainless steels, 81, 84
welds on end grain vs, 85

ENVIRONMENTAL EFFECTS
decisions based on, 46
impurities influence, 38
variations in, 37

ENVIRONMENTS
steel in water and acid, 18
stress corrosion cracking, 115

Equipment, service life factors, 11
Exfoliation causes, 144

F

FABRICATION
copper alloys, component, 255
corrosion fatigue causes, 143

Fasteners, adverse effects from, 167

FATIGUE STRESSES
characterization, 140

coating failure due to, 143
influence of, 140
reduction of, 143
seawater, 142

FIBER-REINFORCED POLYMERS
biological attack vs, 113
pipes, precautions in use, 208
tanks vs corrosive soil, 211

Films and scales, Cu alloys, effects of, 87
Filters erosion-corrosion control, 97
Flame sprayed metal coatings, characterization, 329
Flow, intermittent, consequences of, 97

FUNDAMENTALS
active, passive conditions, 70
bimetallic attack, 67
corrosion cell, 14
design decisions, 17
electrochemical reactions, 15
galvanic series, 68
polarization, 19

G

Galling, fretting causing, 102

GALVANIC CELLS
area effects, 72, 76
atmospheres, test results, 72
avoid in automobiles, 216
carbon, graphite lubricant, 78
coatings to control attack, 77
end grain attack on SS, 82
examples in industry, 76
fundamentals, 67
insulation alternatives and effects, 73, 75

Galvanic systems, cathodic protection, tank protection by, 296

GALVANIZED STEEL
hot water vs, 206
welding, Zn liquid metal attack, 179

Gamma radiation, radiographic tests using, 274

GASKETS
chloride hazards from, 134
materials recommendations, 66

GENERAL CORROSION
statistics on, 24
tabulated data, 47

GEOMETRICAL EFFECTS
heat treatment causing, 190
pitting, 107
plating reduces fretting, 103
welders, space for, 180

Glass coatings, characterization, 331
Grain, causes, cold work, intergranular attack, 144
Graphitization, cast iron, 156
Grit-blasting, fretting damage reduced by, 103

H

Hardness, bolts, surface maxima, 140
Heat affected zone, attack causes, prevention, 150

HEAT EXCHANGERS
anodic protection of, 306
chloride cracking of, 123
design to reduce attack on, 99
maintenance requirements, 12
pitting failure of, 106
rolling tubes into sheets, 182
support designs, 62
temperatures, flow design, 129
test specimens, 30
tube fabrication crevices, 61
tubing vs steam flashing, 101
velocity, other controls, 97
welding tube sheets, 181

Heat flux, test equipment for, 35

HEAT TRANSFER
designs to control, 195
method, influence on rates, 195

HEAT TREATMENT
considerations for, 191
expert advice needed, 192
geometrical problems with, 190
intergranular attack due to, 145
solution, function of, 190
specifications format, 192
SS failure vs, 129
stainless stabilizing, 191
stainless steels vs IGA, 155
tank stress relief, 285
type vs critical, 192
verification, 285
welds, requirements, 188

Heating, local, design to avoid, 195

■ 344 ■

HIGH TEMPERATURES
carbon steel vs, 254
test equipment, 35

Hot dip metal coatings, characterization, 328
Huey tests, acceptance test, 270
HYDROCHLORIC ACIDS
stainless steels vs, 239
titanium vs, 259
velocity influence on attack by, 91

Hydrofluoric acids, titanium vs, 259

HYDROGEN EMBRITTLEMENT
C steel in water, pH level, 253
cracking, characterization, 137
materials vs, 140
mechanism, 137
tantalum and titanium, 123

HYDROGEN GROOVING
carbon steels, 158
tanks and pipelines, 165, 166

Hydrogen, sulfide, cracking due to, 138
Hydrostatic testing, avoidance, reasons for, 114

I
Identification, materials systems, 260, 261, 263, 264

ILLUSTRATIONS
air conditioner tube failure, 106
AISI 410 failure, stress relief, 128
anodic protection design, 307
automobiles, principles, 213, 214, 215
boiler, caustic embrittlement, 117
bronze ground wire in atmosphere, 120
bronze static wire cracks, 121
C steel caustic cracking, 118
cavitation vs pump impellers, 95
chloride cracking of tubes, 131
coatings testers, 322
corrosion data survey characteristics, 40
coupon racks, 287, 288
crevice-free tube sheet, 183
crevices from poor welds, 55
Cu tubing cracking in NH_3, 119
dye penetrant test, cracks, 272, 273
end grain attack, 82-85
excellent weld design, 187
failed Al anodized coating, 326
fatigue failure of pump, 142
fatigue plus corrodent, 141
fluid flow, water box, 100
galvanic series in seawater, 69
good drainage designs, 200
heat exchanger temperature control, 130
heat exchanger test specimen, 30
heat exchanger tube failure, 63
heating coil locations, 196
heavy shell seam weld, 173
HG attack on Al, 80
high-temperature, high-pressure equipment, 36
hydrogen grooving, 160-165
IG attack on SS in HNO_3, 151
impeller design diagram, 66
impingement plates design, 93
inside welds in containers, 54
instantaneous rate meters, 290
insulation alternatives, 76
insulation, area controls, 74
intergranular attack, AISI 316, 146
intergranular attack, 18Cr-8Ni, 145
internal coatings designations, 321
linear polarization meters, 291
materials vs 149 C H_2SO_4, 90
metal ion, oxygen cells, 51
milk cooler failures, 78
mixer impeller failure, 65
Monel corrugated hose SCC, 125
over-rolled tube attacks, 64
pipe internal, external attack, 50
pipeline turbulence effect, 94
pitting, concentration cells, 104
pitting of stainless drum, 105
pump vs sulfuric acid, 89
quench media, 189
radiographic tests of pipe, 275
roof design diagrams, 75
Savannah River plant tanks, 284
seawater vs C steel, 88
sensitized AISI 304 stainless, 147
skip welds, crevices due to, 60
specimen failure in wick tests, 134
specimen racks, 31
SS low C, immune to IGC, 152
stainless steel cracking, 124
stainless tube baffle attack, 64
steel vs H_2SO_4, 19, 158
streamlined piping, 99
structural shape drainage, 199
tank, anodic protection vs H_2SO_4, 308
tank bottom supports, 60
tank cathodic protection, 295, 296, 302
tank corrosion cell, 294

■ 345 ■

tank seam weld designs, 186
tank support design, 203
tanks, rail cars, H grooving, 165
test specimens, 28
thermal insulation effects, 133, 198
tube sheet venting vs Cl, 131
tube-to-tube sheet welds, 185
turbine condensate attack, 96
typical anodic protection curve, 305
velocity test, 33, 34
weld crack radiograph, 280
weld flux failure of Al, 168
weld insert rings, types, 58
weld penetration radiograph, 278
weld porosity radiograph, 281
weld symbol locations, 171
weld types and symbols, 172, 174, 175
weld undercut radiograph, 279
welding C to SS steels, 178
welding qualification test, 176
welding symbols, 169, 170
welding, tanks vs grooving, 166
welding unequal thickness, 320
welds for tube sheets, 182
welds, full penetration, 57
welds vs end grain attack, 85
x-ray inspection of pipe, 276
Zn liquid metal attack on SS, 180

Impellers, for mixers, diagram, design information, 65

IMPINGEMENT, CAUSES
sacrificial plates vs, 93
steam flashing causing, 99
water spray, pitting from, 107

Impressed current, cathodic protection, buried tank, 295
Information systems, consultants, professional, 2

INHIBITORS
anodic, 310
applications, 311
cathodic, 311
characterization, classes, 310
corrosion fatigue vs, 144
deaerators, 311
heat exchangers, use in, 136
identification vs SCC, 136
Na silicate for insulation, 136
organic, 310
oxidizers, 311
sources of information on, 314

vapor phase, 310
water, fresh, 206

Inorganic salts, C steel vs, 254
Insert rings for welds, consumable, 54

INSPECTION, TESTING
equipment, 12, 289
methods selection, lists, 267, 268

INSTANTANEOUS RATE METERS
electrical resistance, 289
high pressure vs, 290
linear polarization, 291
supply sources, 290
temperature limits, 289

INSTRUMENTATION, TEST
coatings, laboratory, 320
heat exchanger, 29
heat flux, 35
high temperature, 35
velocity, 33

Insulation, nonabsorbent vs pitting, 107
Interface effects, films, influence of, 70
Interface, importance of, 315

INTERGRANULAR ATTACK
AISI 300 steels stabilized, 150
alloying with Nb, Ti vs, 150
austenitic stainless steel, 145
carbide formation forming, 145
characterization, 144
controls for, 154
copper alloys, 117
environmental behavior vs, 154
environments causing, 151
heat treatment vs, 155
stabilized SS, economics, 154
stainless steels, low C vs, 150
susceptible metals, 145

Iron sulfates, corrosion rate influence, 251

J

Joining, automobile, correct, 212

L

LABORATORY TESTING
anodic protection, for, 304
materials selection, 27, 29

Leak detection, methods for, 269
Linings, ceramic, glass, description, 331

Linings, coatings, tanks vs hydrogen grooving, 165
Linings, organic, thick film, biological attack, vs, 113
Liquid metal embrittlement, welding, due to zinc, 138

LOCALIZED MATERIALS EFFECTS
chloride concentration, 131
statistics, 46

LUBRICANTS
carbonaceous, attack due to, 78
chloride effects from, 134
fretting damage reduced by, 103

M
Magnetite, sulfuric acid vs, 252

MARINE ATMOSPHERES
Cd, Zn coatings vs, 328
corrosiveness of, 203

MATERIALS SELECTION
database use, 36
organization for, 3

MERCURY
aluminum vs, 80
avoid, 137
copper alloys vs, 117, 256

METAL CLAD METALS
aluminum, 247
sites requiring, 205

METAL COATINGS
anodic and cathodic types, 331
removal vs H cracking, 140

METAL PIPES
cost calculations, 8
steel in seawater, 208
test equipment, 31
tube sheet controls, 182
underground protection, 211
welds, radiographic tests, 274

Metallographic tests, quality controls, 269
Metals, general, metals, prone to fretting, 102
Metals, molten, welds, attack due to, 178
Microstructure, metals, intergranular attack, influence of, 154

Military, U.S. standards, anodized coatings, 324
Mixers, crevice attack in, 63

N
NACE standards and reports, 225
Nickel-chromium-iron alloys, cracking in wet HF + O_2, 124
Nickel-copper alloys, resistance of, 241

NITRIC ACIDS
AISI 304 SS welds vs boiling, 85
aluminum vs, 247
copper vs, 257
stainless steels vs, 238
steels, carbon vs, 249
titanium vs, 259

NITROGENOUS COMPOUNDS
anodic protection vs, 306
copper alloys vs, 117

Nonaqueous liquids not metals, copper vs, 256

NONMETAL PIPES
fiber-reinforced polymer in seawater, 208
polymer, supports for, 209

NONMETALS
chlorides in, 135
fretting control by using, 103

O
OFFSHORE OIL PETROLEUM—WELLS
materials selections, 27, 29
welded specimens, 27

On-site or pilot plant testing, coatings, locations for, 320
Optical testing, weld defects, 277

ORGANIC COATINGS
characterization, 315
specifications, 332

Overlays, repairs using, welded on carbon steel, 329

OXIDATION
causes, fretting, contribution of, 102
passivity due to, 21, 70, 87
titanium, 257

Oxide conversion coatings, characterization, 325

OXIDES
copper vs, 256
rate influence, 38
reactions from, 15
titanium vs, 259

OXYGEN
concentration cell pitting, 104
copper vs, 257
rates influenced by, 206
seawater, influence of, 208

P

PAPER INDUSTRY
anodic protection, used in, 306
roller shell failure, 13

PARTICULATES
air conditioner failure from, 106
controls for polymer pipes, 208
effects of, 38
erosion-corrosion due to, 86
pitting failures due to, 104

PASSIVATION
anodic protection persistence, 304, 308
effect of, 21
oxygen's role in, 21
stainless steel, O_2 for, 311

Peening, corrosion fatigue vs, 143
Petroleum oils, materials vs H_2S in, 141

pH CHANGES
C steel in water, 253
rate influence, 38

Phosphate conversion coatings, characterization, 324
Phosphoric acids, stainless steels vs, 239
Pipelines, grooving in, 165, 166

PITTING
case histories, 104
cathodic protection vs, 107
cleaning to control, 107
mechanism, 103
thickness, increase vs, 107

PITTING CAUSES
austenitic steels, chlorides, 105
bad component geometry, 107

debris, marine growths, 104
sedimentation, 106
standby errors, 101
wet insulation, 106

Plasma spray characterization, 330
Plastic materials, fiber-reinforced pipe, 208
Polarization, mechanism, 19
Polarization rate tests, anodic protection, 304
Polluted atmospheres, surveys to identify, 207
Polyvinylchloride materials, SS cracking by marking bands, 135
Potential-pH diagrams, information in, 39
Potentials, lab testing, how to make, 20
Potentials, underground testing, methodology, 301
Potentiostats, anodic protection, use for, 304
Precipitation hardening steels, description, analyses, 235
Predictive calculations, galvanic cell areas, use for, 72

PRESSURE
bolts for systems at high, 140
instantaneous rate meters vs, 290
pipe, lined vs seawater, 209
polymer pipe limits, 208

Pressure vessels, ASME codes for, 223
Profile, coatings, importance for, 316
Programs, computer, rate data, 39

PUMPS
cavitation effects in, 95
corrosion fatigue vs, 142
materials for seawater, 209
sulfuric acid velocity vs, 88

Pyrophoricity, titanium vs dry chlorine, 260

Q

QUALITY CONTROL
fitness for use decisions, 268
metal tanks, 283
metallographic tests, 269
questions relevant to, 267

R

Railroad tank cars, hydrogen grooving, 165

RATE CALCULATIONS
calculation of, 44
computer programs, 39
maxima recommendations, 47
methods, 42
Nelson charts, 39
nomographs, 42, 45
pitting, precautions for, 107
published bases, 37
steel vs sulfuric acid, 158
sulfuric acid vs C steel, 250
temperature concentration diagram, 39
use of, 46

Rate, pitting, measurement methods, 107

RECORDS, DATABASES
materials selection using, 36
published rates, 37

Reference electrode, cathodic protection, use of, 301
Residual stresses, SS, influence on SCC, 122

S

Safety, specifications, factor in, 229

SEA AQUEOUS
compositional variations, 207
copper vs, 257
fiber-reinforced polymer pipes in, 208
galvanic series in, 68
high pressure vs lined pipe, 209
metal pipes in, 208
oxygen effects in, 208
pump materials vs, 209
stainless pipe in high-pressure, 209
titanium vs, 259

SEGREGATION, GRAIN
intergranular attack due to, 146
stainless steel causes, 150

Shutdown equipment, industry economics, 2
Silicon, cast Cr alloys, contents, 237
Skip welding, effects of, 56
Society of Automotive Engineers, lists of standards published by, 222
Soldering, automobiles, 216

SPECIFICATIONS
cathodic protection, 298
cost factors in, 229
fabrication rules, 218
geometry, specifications, 218
guidelines for writing, 217, 228
heat treatment, 192
proper drainage, 199
safety factors in, 229
standards, use in, 218
surface preparation, 218, 315

SPECIMEN PREPARATION, TEST
crevice corrosion, 28
galvanic corrosion, 28
materials selection, 27
stress corrosion, 28
welded specimens, 27

STAGNANT CONDITIONS
condensate in turbines, 96
reduction by design, 95

NICKEL ALLOYS
alkalis and salts vs, 239
Alloys Hastelloy B-2, C-276, G-3, 244
Alloys 201 and 400, 241
Alloys 600 and 625, 242
Alloys 800, Carpenter 20 Cb-3 825, 243
Hastelloy G-30, C-22, 245
properties vs corrosion, 239
resistance of, 241
SCC, environment vs failures, 123

STANDARDS
company compilations, 228
how to use, 220
organizations writing, 219
publishers of, 218
specifications, use in, 221
use of national, 218

STANDBY TECHNIQUES
equipment handling water, 101
pitting, solution controls, 107
steam turbine, 96

STATISTICS
corrosion failures, general attack, 24
localized attack, 46
stress corrosion cracking, 24

STEAM
C steel vs, 254
flashing, impingement by, 99
flashing vs heat exchanger tubes, 101

■ 349 ■

Steam turbines, steam flashing vs blades, 99

STEELS CARBON
compositions, uses, 248
SCC, environment vs failures, 123
weathering, atmospheric effects on, 253

Steels, cast Cr-Ni, analyses, 235, 237

STEELS, CHROMIUM-NICKEL
acids vs, 238, 239
alloying and types of, 231
austenitic identification, 232, 234
chloride stress cracking, 122
descriptions, analyses, 235
end grain attack in HNO_3, 84, 85
knife line attack, 150
martensitic identification, 233
nitric acid resistance, 238
stabilized grades, 150
utility, development, 230
velocity limits in seawater, 209
wrought, compositions, 236

Steels, economics, 1
Steels, nickel, chlorides vs, 124
Strauss test, materials qualification by, 271

STRESS CORROSION CRACKING
austenitic SS, temperature influence, 122
characterization, 115
chlorides vs austenitic SS, 122
copper, alloying influence vs, 120
copper, season cracking, 117
environments causing, 115
inhibitors vs, 136
metals selection vs, 126
shot peening vs, 135
steels, carbon, 115
temperature controls vs, 130

Stress relieving, characterization, 190

STRESSES
bronze grounding wire vs, 119
cold work, due to, 189
hydrogen embrittlement, influence of, 138
internal, failures due to, 129
internal, reduction vs SCC, 126
SS austenitic influence on SCC, 122
tight fits causing high, 129

Sulfates, graphitization by, 156

SULFURIC ACIDS
anodic protection C steel vs, 253, 306
C steel, attack rate, 158
contaminated, velocity influence, 91
corrosion allowance, 253
dilution controls, 251
pipe, velocity tests in, 89
pumps vs, 88
titanium vs, 259

SURFACE EFFECTS
fretting due to finish, 102
smoothness vs cavitation, 94

SURVEILLANCE ON SITE
coating systems, 205
comprehensive review, 267
equipment, 12
inhibitor systems, 314
monitoring applications, 292
necessity for continuing, 286
thickness tests, 289
water systems, 314

T

TABULATED DATA
austenitic stainless steels, 232
C steel sttress relief, 191
cast stainless steel analyses, 237
copper alloy compositions, 255
galvanic series in atmosphere, 70
galvanic series in plants, 71
industrial coatings data, 318
inhibitors vs corrosives, 312
liquid metal cracking combinations, 139
low carbon steel uses, 248
martensitic stainless steel, 233
materials system example, 265
Ni alloys analyses, uses, 240
Ni-Cr SCC, Mg Cl, time factors, 126
SCC alloy-environmental systems, 116
solutions causing SS IGC, 153
stainless steel finishes, 219
stainless steels, wrought, 236
stress corrosion cracking statistics, 123
titanium properties, 258
UNS designations, 261, 262

TANKS, METAL
aluminum alloys, 13
anodic protection on, 306
bottom failure case histories, 56, 59, 60
cathodic protection of buried, 294, 295
crevice attack on, 56

■ 350 ■

drainage design, 199
horizontal, supports for, 60
hydrogen grooving controls, 163, 164
leaking, pollution by, 210
pouring lip for, 202
quality controls for, 283
support design for drainage, 200
supports, protection of, 202

Tantalum, hydrogen embrittlement, 123

TEMPERATURES
Al vs organic acids, influence, 247
austenitic SS, influence on SCC, 122
caustic failure influence, 118
cold areas effects, 197
H embrittlement factor, 138
hot wall effects, 194
instantaneous rate meters, 289
lowering, effects of, 129
Ni-base alloys in caustics, 124
reaction kinetics, 193
stainless chloride cracking, 124
tanks, service life estimates, 211
titanium limits, 259
water, importance of, 313

Tensile stresses, avoidance of, 135

TESTING
acceptance, methods, 270
Al alloys, heat treatable, 271
biocides, 113
boiling HNO_3 for SS, 154
boiling HNO_3 test for IGC, 154
coatings, 319
cracks, 271
crevice, 271
eddy current, 282
heat transfer effects, 195
heat treatment verification, 285
hydrostatic, dangers of, 114
intergranular attack susceptibility, 153
materials for water system, 207
metals vs seawater, 207
polarization curves uses, 20
procedures, solutions, 32
prospective plant sites, 204
radiographic, 272
soil corrosiveness, 210
sulfuric acid velocity, 89
tanks, 284
time factor in inspections, 230
weld radiography inspection, 229
welder qualification, 283

welding qualification, 176
wick test for Cl concentration, 133

THERMAL INSULATION
benefits from proper, 198
chloride concentration under, 132
coatings on SS vs SCC, 136

TIME EFFECTS
inspection scheduling, 230
plant life factor, 45

TITANIUM
cleaning for high temperature, 137
hydrogen embrittlement, 123

Topcoats, galvanizing, advantages of, 329

TURBULENCE
pipe design to reduce, 98
pipeline effects, 94
valves producing low, 98

U

Ultrasonic testing, characterization, standards, 277, 282
Underground environments, corrosiveness variations, 210

V

Valves, diaphragm, low turbulence, 98

VELOCITY
acids increase effects of, 88
C steel in seawater, influence, 88
C steel vs H_2SO_4, influence of, 251
concentration cell influence, 95
Cu alloys, effects on, 87
effects due to turbulence, 94
HCl attack, influence of, 91
H_2S in pipelines, 166
impingement plates vs, 93
minimum requirements, 95
pipe size change economics, 92
pumps vs sulfuric acid, 88
reduction in heat exchangers, 97
seawater vs stainless steel, 209
sulfuric acid vs pipes, 89
water, importance of, 313
water limit, polymer pipes, 208

■ 351 ■

W

WATER
acidic, polluted, controls, 206
aluminum vs, 247
analysis, 206
biocide use in, 113
carbon steel vs, 253
cleaning importance, 314
composition, importance of, 313
dealloying propensities, 157
galvanized steel in hot, 206
industries, economics, 1
materials choice, variables in, 313
sterilization methods, 114
supply change vs MIC, 113
systems, choices among, 313

Weight loss, on-site cathodic protection surveillance, 301

WELDED JOINTS
clad metals, techniques, 330
defects shown radiographically, 275
radiographic tests, 275
unequal thickness, 320

WELDING
adverse consequences of, 167
AISI 304 failure from sensitization, 147
backup rings diagrams, 58
consumable insert rings, 54
copper alloys, 254
corrosion fatigue vs, 143
crevices due to, 51, 60, 170
design precautions, 167
electrode selection vs IGC, 154
excellent designs, 186
filler metal selection, 13, 175
flux failure of Al, 168
geometrical considerations, 180
heat treatment effects, 187, 188
liquid metal attack from, 178
materials compatibility, 177
metal coatings manual and automatic, 329
pipe, radiographic tests, 274
qualification of operators, 176, 283
slag inclusions from, 173
sulfuric acid, quality vs, 252
symbols, 51, 168, 169, 171, 174, 175
tanks vs H grooving, 166
tests, safety factors, 229
thin wall stainless pipe, 179
titanium, 257
tube sheet controls, 181, 184
type advantages, disadvantages, 173
wall dilution controls, 177
weld heavy seam and joint, 173

Wire, electrical, Al coated steel in atmosphere, 119

X

X-ray radiographic tests, 274

Z

ZINC
alloys for industry, 256
brackish and seawater vs, 259
characterization, 256
chlorine, wet vs, 260
coatings, 328
hydrochloric, H_2SO_4 vs, 259
hydrofluoric acid vs, 259
industrial applications, 256
inhibition in oxidizing solutions, 259
nitric acid vs, 259
oxidizing salts vs, 259
welding procedures, 256